Connecting the NCTM Process Standards **and** the CCSSM Practices

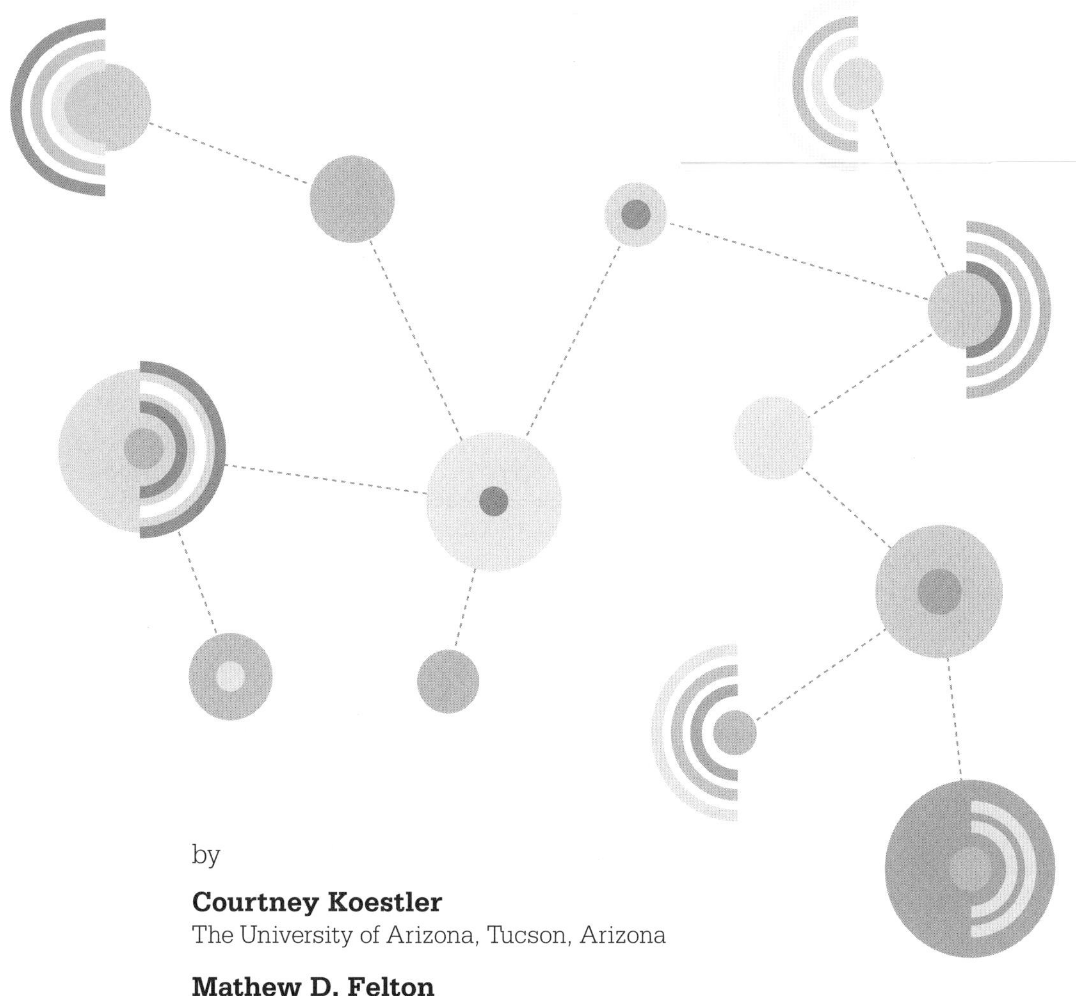

by

Courtney Koestler
The University of Arizona, Tucson, Arizona

Mathew D. Felton
The University of Arizona, Tucson, Arizona

Kristen N. Bieda
Michigan State University, East Lansing, Michigan

Samuel Otten
University of Missouri, Columbia, Missouri

Cataloging-in-Publication Data is on file with the Library of Congress.

ISBN 978-0-87353-708-7

The National Council of Teachers of Mathematics is the public voice of mathematics education, supporting teachers to ensure equitable mathematics learning of the highest quality for all students through vision, leadership, professional development, and research.

Printed in the United States of America

TABLE OF CONTENTS

FOREWORD

The Common Core State Standards Initiative (CCSSI) may be the most profound and widely distributed educational reform activity in recent history. One of the results of its efforts, the *Common Core State Standards for Mathematics* (CCSSM 2010), will affect nearly every K–12 student and the majority of the nation's teachers over the next decade. CCSSI focuses on a foundational element of K–12 education—curriculum—as a means to improve student learning opportunities. Although CCSSI was formed using a top-down approach at the prompting of a group of governors and chief state school officers, the primary audience for and ultimate users of the standards are classroom teachers.

The question of whether CCSSM represents the best thinking in the field or whether the existence of common standards, in general, will advance or inhibit innovative curriculum development and evaluation is unresolved. However, what is clear is the need to support teachers in their work helping students achieve a high-quality mathematics education. Given the centrality of curriculum in the work of teachers (and students), supporting teachers as they prepare, plan, and implement instruction that attends to shared educational goals is critical. The adoption of CCSSM by 45 states and the District of Columbia represents a watershed moment of consensus and an opportunity to support teachers in a common goal.

The primary focus of this volume is on elaborating the Standards for Mathematical Practice outlined in CCSSM. These standards are given front-page attention in the CCSSM (that is, they are introduced early in the document and intended to permeate the entire K–12 curriculum). However, the brevity with which they are described and the lack of integration into the standards for mathematical content make them easy to overlook or ignore. Doing so will substantially dilute the impact of and potential for CCSSM to improve student learning of mathematics.

The ideas in the Standards for Mathematical Practice are not new but rather linked to mathematical goals articulated in previous documents and by other groups. In fact, problem solving and reasoning are at the core of all the practices outlined in CCSSM, as they have been at the core of the work of the National Council of Teachers of Mathematics (NCTM) since the publication of *An Agenda for Action* in 1980. In fact, the first recommendation of the *Agenda* was that "Problem solving must be the focus of school mathematics" (p. 2). The document went on to say, "Performance in problem solving will measure the effectiveness of our personal and national possession of mathematical competence" (p. 2).

In subsequent NCTM curriculum recommendations—*Curriculum and Evaluation Standards for School Mathematics* (1989) and *Principles and Standards for School Mathematics* (2000)—the role and place of mathematical processes and practices are underscored and further elaborated. There are other instantiations where these ideas have been eloquently presented. For example, in 1996, Cuoco, Goldenberg, and Mark wrote about the importance of mathematical habits of mind as a central focus of mathematics instruction:

> Although it is necessary to infuse courses and curricula with modern content, what is even more important is to give students the tools they will need in order to use, understand, and even make mathematics that does not yet exist. A curriculum organized around habits of mind tries to close the gap between what the users and makers of mathematics do and what they say. (p. 376)

Likewise, the authors of *Adding It Up: Helping Children Learn Mathematics* (NRC 2001) highlighted the need to think more broadly about what it means to learn mathematics:

> Recognizing that no term captures completely all aspects of expertise, competence, knowledge, and facility in mathematics, we have chosen mathematical proficiency to capture what we believe is necessary for anyone to learn mathematics successfully....Mathematical proficiency, as we see it, has five components, or strands: conceptual understanding, procedural fluency, strategic competence, adaptive reasoning, productive disposition. (p. 116)

The Standards for Mathematical Practice outlined in CCSSM, and discussed more fully in this volume, are a reaffirmation of the significance of and a call for renewed emphasis on these same themes (habits of mind, mathematical processes, mathematical proficiency) as a significant aspect of learning mathematics. While some terms may be new to teachers, the central ideas have been around a long time and remain the same. The authors of this volume recognize and make explicit connections between these related ideas and the CCSSM Standards for Mathematical Practice.

A major contribution of this volume is the attention to making explicit, through classroom vignettes, images of the mathematical practices. These vignettes come from classrooms at different grade levels (elementary, middle, and secondary) and within the context of different mathematical content. This type of elaboration is essential in helping users of CCSSM (e.g., teachers, curriculum and assessment developers, teacher educators) apply and connect the practices to content and classroom instruction.

In the 1980 *An Agenda for Action*, the NCTM Board of Directors acknowledged its professional obligation:

> We recognize as valid and legitimate the role of public opinion in the determination of educational goals. But this philosophy is predicated on a well-informed public. Thus, the National Council of Teachers of Mathematics, as an organization of professional educators, has a special obligation to present its responsible and knowledgeable viewpoint of the directions mathematics programs should be taking in the 1980s. (p. i)

With this NCTM volume, the authors continue a tradition of supporting teachers in their enactment of a core curriculum that is based on this vision.

Barbara J. Reys
Curator's Professor and
Lois Knowles Faculty Fellow
University of Missouri—Columbia

References

Common Core State Standards Initiative (CCSSI). *Common Core State Standards for Mathematics.* Washington, D.C.: National Governors Association Center for Best Practices and the Council of Chief State School Officers, 2010. http://www.corestandards.org.

Cuoco, A., E. Goldenberg, and J. Mark. "Habits of Mind: An Organizing Principle for Mathematics Curricula." *Journal of Mathematical Behavior* 15, no. 4 (1996): 375–402.

National Council of Teachers of Mathematics (NCTM). *An Agenda for Action.* Reston, Va: NCTM, 1980.

National Council of Teachers of Mathematics (NCTM). *Curriculum and Evaluation Standards for School Mathematics.* Reston, Va: NCTM, 1989.

National Council of Teachers of Mathematics (NCTM). *Principles and Standards for School Mathematics.* Reston, Va: NCTM, 2000.

National Research Council (NRC). *Adding It Up: Helping Children Learn Mathematics.* Edited by J. Kilpatrick, J. Swafford and B. Findell. Washington, D.C.: National Academies Press, 2001.

INTRODUCTION

> Working at teaching turns out to be something like navigating in a multidimensional terrain, getting safely across the street while also crossing the city, sighting one island from another while catching the wind that makes it possible to get around the corner and across the ocean. (Lampert 2001, p. 50)

Teaching mathematics is an incredibly complex task. Teachers must help students learn mathematical content while also guiding them in what it means to *do mathematics.* The eight Standards for Mathematical Practice in the new *Common Core Standards for School Mathematics* (CCSSI 2010) provide one vision of doing mathematics, and this book is intended as a roadmap to help teachers navigate these practices.

The *Common Core State Standards for Mathematics* (CCSSM) were developed as part of a project sponsored by the National Governors Association and the Council of Chief State School Officers in an effort to "define what students should understand and be able to do in their study of mathematics" (CCSSI 2010, p. 1). Although the authors of this document assert that the CCSSM were not intended to be national standards, 45 states and the District of Columbia have formally adopted or endorsed the standards. Only 5 states have not adopted the standards at this time.

While grade-specific content standards make up the body of CCSSM, it begins with the eight Standards for Mathematical Practice. Although the description of each practice is brief in the CCSSM document, the mathematical practices describe what it means to do mathematics and should therefore permeate mathematics instruction across grade levels and content areas. Careful integration of the mathematical practices into classroom instruction is of particular importance because mathematics content is often foregrounded (and assessed) over mathematical practice and process.

The mathematical practices are written as different skills, dispositions, and understandings of mathematics that students should have; however, it is teachers who must provide meaningful experiences in which students may develop these various forms of expertise. This book is designed to help in that endeavor by elaborating on each of the eight practices: describing each practice in more detail by discussing connections to the National Council of Teachers of Mathematics' (NCTM) five Process Standards (NCTM 2000), providing elementary and secondary classroom examples of the practices, and giving readers additional resources to seek out for each practice. Similar to the mathematical practices themselves, each chapter is slightly different in terms of structure, scope, and size. While they are presented in the CCSSM as individual practices, we believe it is important for teachers to integrate the practices in their pedagogy. Therefore, the final chapter of this book highlights some of the important connections among practices.

Note: While the authors are listed in order as they joined the project, all four authors contributed equally in the development and writing of this book. The authors wish to thank Dr. Jack Smith and Dr. Joel Amidon for reviewing our work and providing feedback and Janet M. Liston for contributing to the high school vignette section of practice 1.

NCTM Resources

While each chapter lists additional resources, most from NCTM, that are relevant to the specific mathematical practice, there are two important sets of NCTM resources that span all the practices, and we strongly encourage you to read them to deepen your understanding of the mathematical practices.

- NCTM's Navigations series is an excellent resource to support teachers in infusing their teaching with the *Principles and Standards for School Mathematics* (NCTM 2000). Each volume is geared toward a specific grade level or grade band and focuses on one or more aspects of the *Principles and Standards*. The series is full of excellent tasks that teachers can use with their students to support their learning with understanding. Each book also comes with a CD-ROM containing activities to use with students, printable activity pages, articles from NCTM's journals, and interactive applets. See the NCTM catalog online at http://www.nctm.org.
- NCTM's Essential Understandings series is another excellent resource. Each book in the series provides a comprehensive overview of the big ideas and essential understandings of mathematical topics in the K–12 curriculum. Currently eleven titles are available, with titles including *Developing Essential Understanding of Expressions, Equations, and Functions for Teaching Mathematics in Grades 6–8* and *Developing Essential Understanding of Multiplication and Division for Teaching Grades 3–5*, with six more titles forthcoming. These books are a helpful tool in learning about the conceptual structures of the ideas in each topic and are useful in planning lessons that assist students in recognizing important mathematical concepts and structures.

References

Common Core State Standards Initiative (CCSSI). *Common Core State Standards for Mathematics.* Washington, D.C.: National Governors Association Center for Best Practices and the Council of Chief State School Officers, 2010. http://www.corestandards.org.

Lampert, M. *Teaching Problems and the Problems of Teaching.* New Haven, Conn.: Yale University Press, 2001.

National Council of Teachers of Mathematics (NCTM). *Principles and Standards for School Mathematics.* Reston, Va: NCTM, 2000.

Make Sense of Problems and Persevere in Solving Them

Practice 1: Make sense of problems and persevere in solving them

Mathematically proficient students start by explaining to themselves the meaning of a problem and looking for entry points to its solution. They analyze givens, constraints, relationships, and goals. They make conjectures about the form and meaning of the solution and plan a solution pathway rather than simply jumping into a solution attempt. They consider analogous problems and try special cases and simpler forms of the original problem in order to gain insight into its solution. They monitor and evaluate their progress and change course if necessary. Older students might, depending on the context of the problem, transform algebraic expressions or change the viewing window on their graphing calculator to get the information they need. Mathematically proficient students can explain correspondences between equations, verbal descriptions, tables, and graphs or draw diagrams of important features and relationships, graph data, and search for regularity or trends. Younger students might rely on using concrete objects or pictures to help conceptualize and solve a problem. Mathematically proficient students check their answers to problems using a different method, and they continually ask themselves, "Does this make sense?" They can understand the approaches of others to solving complex problems and identify correspondences between different approaches. (CCSSI 2010, p. 6)

Unpacking the Practice

The first mathematical principle in the *Common Core State Standards for Mathematics* (CCSSM; CCSSI 2010) centers on *problem solving*—making sense of problems and persevering in solving them. In this section, we show how this practice is aligned with each of the National

Council of Teachers of Mathematics Process Standards (NCTM 2000). In addition, because problem solving is so fundamental to learning mathematics with understanding, we briefly mention how certain aspects of this mathematical practice are explored further in later chapters focused on the other mathematical practices.

Problem Solving Standard

Although problem solving is just one of the Process Standards in NCTM's (2000) *Principles and Standards for School Mathematics*, it is central to the kind of mathematics learning NCTM advocates. Because problem solving is part of all content areas, problem-solving activities should not be an isolated part of a lesson, unit, or curriculum but should instead be integrated into students' experiences, involve important mathematics, and connect to multiple process and content strands.

"Problem solving means engaging in a task for which the solution method is not known in advance. In order to find a solution, students must draw on their knowledge, and through this process, they will often develop new mathematical understandings" (NCTM 2000, p. 52). This means that many of the tasks we ask students to engage in must be problems to solve, not simply exercises to do. These tasks should allow students to enter problems via multiple entry points and to invent and use strategies that make sense to them. It is through this meaning making that students can develop and deepen their mathematical understanding.

If we want students to develop what *Adding It Up* (NRC 2001) calls a *productive disposition* as a mathematical problem solver, we must support them in taking up a different role than typically seen in traditional classrooms. *Adding It Up* defines a productive disposition as "the tendency to see sense in mathematics, to perceive it as both useful and worthwhile, to believe that the steady effort in learning mathematics pays off, and to see oneself as an effective learner and doer of mathematics" (p. 131). Students should be active participants in the sense-making process, responsible for making sense of the problems before them and allowed to invent solution strategies by building on the knowledge they already possess. We know from the work of Carpenter and colleagues (1999) that even young children can solve many different kinds of problems—without having to be explicitly taught—by using their own informal mathematics knowledge; in fact, "children may actually understand the concepts that we are trying to teach but be unable to make sense of specific procedures that we are asking them to use" (Carpenter et al. 1999, p. xiv). When teachers allow students to use and build on their own knowledge, students can strengthen and extend what they know as well as develop new mathematical understanding when problem solving.

The teacher plays a crucial role supporting students in problem solving. Although many teachers have textbooks to guide their instruction, it is their responsibility to select, adapt, design, and implement appropriate mathematical tasks for the students in front of them. If mathematical tasks are posed to students in ways that allow them to build on what they already know, students are more able to develop their own methods for solving problems as well as create new knowledge in the process (NCTM 2000). Teachers should create classrooms where students are encouraged to "explore, take risks, share failures and successes, and question one another...[so that] they will be more likely to pose problems and to persist with challenging problems" (NCTM 2000, p. 52), just as this mathematical practice calls for.

Reasoning and Proof Standard

The Reasoning and Proof Standard makes brief mention of the importance of making conjectures to assist in solving a problem (NCTM 2000, p. 55). Students should have opportunities to make and communicate conjectures, to explore these conjectures, and to analyze and justify these conjectures. Making and investigating conjectures supports students in developing knowledge that can be used to understand a problem, see how a problem is connected to a larger body of mathematics, and learn new mathematics. However, because this idea is more fully developed in practice 3: construct viable arguments and critique the reasoning of others, please see that chapter for more detail.

Representations Standard

Although practice 1 states that "*younger* students might rely on using concrete objects or pictures to help conceptualize and solve a problem" (CCSSI 2010, p. 6; italics added), students of all ages can benefit from using manipulatives or drawing pictures to help understand and solve problems. By using concrete objects or pictures, students can understand the underlying concepts in a problem and may be able to make connections to more abstract ways of representing the problem.

NCTM's *Principles and Standards* describes several ways students should learn to work with representations in pre-K–12 classrooms: "create and use representations...; select, apply, and translate among representations...; use representations to model...phenomena" (NCTM 2000, p. 67).

Teachers should provide opportunities for students to learn about and use conventional representations as well as opportunities for students to invent and develop their own ways of representing mathematical ideas and thinking. In lower grade classrooms, this might simply be encouraging students to draw pictures as a way to solve a problem. In upper grade classrooms, students might "transform algebraic expressions or change the viewing window on their graphing calculator to get the information they need," as practice 1 states. Students at all grade levels should be supported in developing their own representations and challenged to explain the connections between their representations and the problem itself, as well as the connections among representations. Finally, students should have opportunities to model various phenomena using appropriate representations, something that is explored in more detail in the chapter on practice 4: model with mathematics.

Connections Standard

NCTM, in its *Principles and Standards,* states that students should "recognize and use connections among mathematical ideas" (2000, p. 64) that can support them in "consider[ing] analogous problems, and try[ing] special cases and simpler forms of the original problem in order to gain insight into its solution" (practice 1; CCSSI 2010, p. 6). If students can do this, they are better able to develop a richer and deeper understanding of the problem at hand and to understand the discipline of mathematics as an integrated body of knowledge rather than discrete, unrelated topics.

NCTM's Connections Standard says that students should have opportunities to recognize and use mathematics in contexts outside of mathematics (2000, p. 64). Although this idea is not an explicit part of CCSSM's practice 1, it is closely connected. Teachers can broaden their students' understanding of mathematics by connecting to other school subjects like science and social studies. Teachers can also engage students in understanding mathematics more deeply and personally by connecting to, including, and building on students' *community mathematics knowledge* (Civil 2007; Gutstein 2006, 2007). Community mathematics knowledge is the mathematical knowledge that students can gain and use in their out-of-school experiences. By grounding problem solving in students' experiences and focusing on the assets that students bring to the classroom, teachers can highlight the important mathematical practices in which students and their families engage, as well as the relevance of mathematics to their lives. This doesn't mean that all tasks need to be set in real-world contexts. Abstract tasks that develop, support, and challenge students' understanding of and connections among mathematical content and processes are also important.

Communication Standard

Practice 1 ends with a statement about the need for students to understand others' solution strategies as well as understand the mathematical connections between different strategies, which aligns with important aspects of NCTM's Communication Standard. In fact in meeting the Standard, not only will students learn to understand and evaluate others' strategies, but when they engage in mathematical arguments where they must justify their own solutions, they "will gain a better mathematical understanding as they work to convince their peers about different points of view" (NCTM 2000, p. 60, citing Hatano and Inagaki 1991).

Mathematical discussions should center on ideas and solution strategies so as to place attention on mathematical understanding rather than simply solutions (Hiebert et al. 1997). It is important to note that these ideas and strategies should, for the most part, come from the students themselves. Teacher and students should have joint responsibility for sharing and clarifying important mathematical ideas. This does not mean that teachers should be passive observers waiting for their students to "discover" the mathematics, but rather that teachers must play an important role in discussions in which students have opportunities to explain ideas, to show solution strategies, and to determine the correctness of an answer themselves. When students develop their own methods, their opportunity to learn mathematics with understanding increases (Carpenter and Lehrer 1999; Hiebert et al. 1997).

Classroom Examples

In this section, we present three classroom vignettes. Although there are many different ways that teachers can encourage students to make sense of problems and persevere in solving them, we provide one elementary grades vignette and two secondary grades vignettes to illustrate several important aspects of the practice.

Elementary Grades Vignette: How Many Teams?

To engage young students in problem solving, teachers can and should use problems that come from the students' worlds (NCTM 2000). Although some elementary school teachers may be reluctant to spend the majority of their time on story problems because they fear that they are too difficult or that students must be able to recall number facts first, students in classrooms with a strong focus on problem solving perform just as well as students in classrooms that have a more traditional focus on only number facts (e.g., see Peterson et al. 1989).

In the following vignette, a teacher poses a division problem arising from an event taking place in his class. Although the teacher has not formally introduced division to his second graders, he is confident that they will be able to solve the problem by using strategies that make sense to them.

Mr. Guzman poses the following problem to his students: "As you know, we are joining Ms. Leavell's class to play math games this afternoon. If there are 36 kids and we need to make teams of 4, how many teams will there be?" The teacher allows several minutes for the students to explore the problem individually using various strategies, then asks students to discuss their strategies in small groups. Finally, students are asked to share their solution methods with the whole class.

Mr. Guzman:	Who would like to explain how they solved the problem to the class?
Suzanna:	I can. I drew a picture. All I had to do was draw 36 dots, and then I drew circles around groups of 4.
Mr. Guzman:	Why did you use 36 dots?
Suzanna:	I used 36 dots because there were 36 kids. I didn't want to draw actual people...it would take too long, so I just used dots.
Mr. Guzman:	Did anyone else use a drawing to help them solve the problem?
	[*A few students raise their hands, and Mr. Guzman asks the students to briefly discuss how the drawings are similar to or different from Suzanna's and from one another's.*]
Mr. Guzman:	Did anyone use another strategy?
Katy:	Yeah, I used blocks.
Mr. Guzman:	How did you use blocks to help you solve this problem?
Katy:	Well, first I counted out 36 blocks to stand for each kid. Then I just started making them up into groups of 4 to make the teams.
Mr. Guzman:	And then what?
Katy:	Then I just counted up how many groups I made.

[*Again Mr. Guzman asks if any other students used manipulatives and very briefly discusses how the manipulatives were used in similar or different ways and even connects them back to the drawings previously discussed.*]

Mr. Guzman: Are there any other strategies?

Carlos: I just started adding 4s.

Mr. Guzman: OK, how did you know to start adding 4s?

Carlos: Because I knew there would be 4 people on each team.

Mr. Guzman: OK, so when you added 4 and 4, that would be 8 people or 2 teams. Then you added another 4?

Carlos: Yep.

Mr. Guzman: OK, so then what? How did you know how many 4s to add?

Carlos: Well, I just kept adding, like 4, 8, 12, until I got to 36 because that would mean that I included all 36 kids. To figure out the answer, I then just had to count how many 4s I added up.

Mr. Guzman: And what did each 4 stand for?

Carlos: One team.

Mr. Guzman: So what was your answer? How many teams will there be?

Carlos: 9.

Mr. Guzman asks Suzanna, Katy, and the rest of the class if they also got 9 teams as the answer and then begins a discussion about the mathematical connections among the solution strategies (e.g., when Suzanna circled groups of 4 dots this was similar to Katy dividing up the blocks in groups of 4).

Notice that when Mr. Guzman asked the students about how they solved the problem, the main focus was on students' problem-solving strategies, not the answer. While the answer is important, Mr. Guzman wanted his students to focus on their own solution strategies and then to consider mathematical connections among these strategies in this mathematical discussion.

Although the problem in this example is a partitive division problem and may at first seem difficult for primary students, it affords multiple entry points, which allow students to use different kinds of strategies (e.g., drawing a picture, using manipulatives, using addition) and shows that young children can successfully solve this problem using various strategies (Carpenter et al. 1999). This vignette illustrates how students who have never been formally taught division can still solve the problem in ways that make sense to them. By having tasks that include multiple solution paths, students are challenged not only to make sense of the problem but also to understand it more deeply by reflecting on the mathematical connections among the different strategies

Middle School Vignette: The Border Problem

This example features the Border Problem and shows how a teacher might introduce middle school students to algebraic thinking and representation (see Burns and Humphreys [1990] and Boaler and Humphreys [2005] for more detailed information and analysis of teacher moves and student strategies). It is similar to the elementary-grades example in that the teacher selects a problem-solving task that has multiple entry points for students, encourages them to make sense of different solution strategies, and provides ways for students to see the connections among strategies.

Ms. Waters:	For today's problem, I'm going to show you a picture on the overhead. It's a 10 x 10 grid, and some of the squares are colored in. What I want you to figure out is how many of the squares are shaded in on the 10 x 10 square and to do it without counting each of the squares individually. I'd like you to try to come up with more than one strategy so that you can check your answer, since I don't want you to count them.
	[*Ms. Waters then shows figure 1.1 and gives them a short time to think.*]

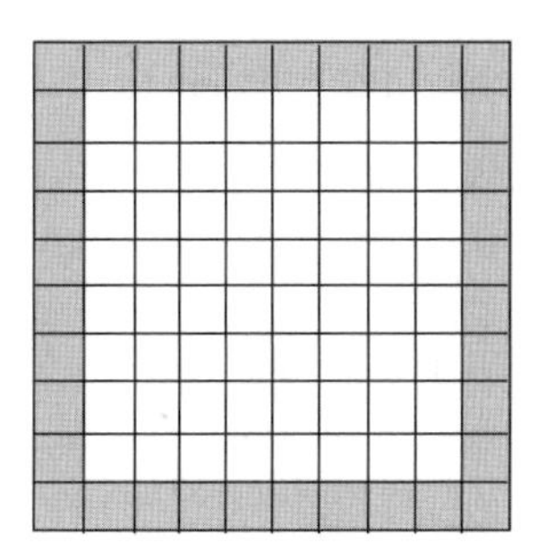

Fig. 1.1. The Border Problem

Ms. Waters:	I bet you might have a few ideas of how to figure this out, but first I want to hear what you think the answer might be. How many are shaded in?
Larissa:	40!
Finn:	36!
Johnetta:	38?
Ms. Waters:	Hmmm, so right now we have 40 and 36 and 38 as possible answers on the table. Did anyone else get another answer?
	[*No one offers another answer.*]

Ms. Waters:	OK, so do you think that this is the kind of problem that could have more than one answer?
Larissa:	No, because we're all counting the same squares. There should be only one right answer.
Ms. Waters:	OK, so take some more time to continue working. Be sure to come up with a strategy that you think will help you arrive at the right answer. Remember, I also want you to challenge yourself to come up with multiple solution strategies. You can also work with people at your table.
	[*After several minutes, Ms. Waters asks students to share their strategies with the class, and these get recorded on the board for the students to see.*]
Ms. Waters:	Finn, can you share your strategy?
Finn:	I think there are 36 colored squares.
Ms. Waters:	OK, why? How did you get that answer?
Finn:	I got this because I added 10 for the top row, 10 for the bottom row, and 8 for each of the columns on the right and left sides.
Ms. Waters:	Tell me more. How did you decide that there were 8 on each side?
Finn:	Well, at first I thought there were 40 squares shaded in…
Larissa:	Me too!
Finn:	But then I realized that I was counting some of the squares more than once—the squares on the corners.
Larissa:	I also thought there were 40 squares, but I realized that wasn't right because that would mean there were 60 squares that weren't shaded. And I knew that wasn't right.
Ms. Waters:	How did you know that wasn't right?
Larissa:	When we had time to work with people at our table, Joslyn's strategy made me realize this.
Ms. Waters:	Joslyn, can you tell us what you did?
Joslyn:	Well, I wasn't sure this was going to work at first, but I decided that instead of counting the shaded squares, I could figure out how many were not shaded and then subtract that number from 100.
Ms. Waters:	You subtracted it from 100?
Joslyn:	Yes, because you told us that it was a 10 x 10 grid, which means there were 100 squares total. I could visualize the 8 x 8 square in the middle, which would be 64. 100 minus 64 is 36.

Ms. Waters: You said that you visualized 64. Were you sure it was 64?

Joslyn: Yes, because one row across the whole square would be 10, but there is a square shaded on both ends which makes the inside square only 8 across.

Ms. Waters: Did anyone else try this strategy?

[*Only one other student raises his hand.*]

Ms. Waters: Ok, so take a moment, and talk with someone close by about why this might be a good strategy.

After a short time, Ms. Waters asks students to share their strategies. Not only are these strategies discussed and recorded, the teacher and students discuss the mathematical connections among the strategies. Ms. Waters will follow up by using different-sized square grids to support students in beginning to generalize their solutions for n by n grids.

In this vignette, the students are engaged in the problem not because it is from a context taken from their everyday lives, but because it is a problem that has multiple entry points students can use to solve it. Although at first the students arrive at different answers, it is clear that the students are able to persevere in figuring out the correct answer. The students also begin to discuss the connections between the diagram and their strategies, eventually connecting to generalized equations that would allow them to solve related problems.

High School Vignette: Completing the Square

This final vignette features a high school teacher inspired by Vinogradova's (2007) description of a lesson introducing students to "completing the square." When completing the square is introduced to students, it is most often explained from a procedural perspective because many students will not use this method in actual practice when solving quadratic equations unless they are asked to. The completing the square method (CTS) becomes important as students learn to solve quadratic equations and use translation methods to graph functions (and conics). Additionally, CTS is used in integral calculus.

This activity, however, supports students in understanding the underlying concepts through making connections between the algebraic procedure of completing the square and its geometric meaning. The following vignette involves students in an algebra II class; they have just finished sections on graphing quadratic equations from tables of values. At this point, the students know how x-intercepts connect to solutions of quadratic equations and have some knowledge of squares and experience with algebra tiles.

To begin the activity, students are asked to use algebra tiles to square binomials, for example $(x + 3)^2$. The common error students make is to square only the x and the 3, yielding an incorrect answer of $x^2 + 9$. Since this activity has been done before, students working in groups are able to build the correct representation of $x^2 + 6x + 9$ with their tiles. Note: the algebra tiles consist of x-by-x squares, 1-by-x rods, and 1-by-1 small squares (unit squares). The teacher poses a question to the class.

Ms. Liston: Would you be able to work "backward"? By that I mean, if you started with the trinomial $x^2 + 6x + 9$, would you be able to write the binomial you would square to get that trinomial?

Sam: Sure, $(x + 3)^2$

Ms. Liston: Right. Would you be willing to try another one?

[*Sam nods.*]

Ms. Liston: How about $x^2 + 8x + 16$? I'd like all the groups to work on this while Sam is thinking about his answer. Sam, feel free to talk with your group.

Sam: I got $(x + 4)^2$.

Zoe: I got $(x + 8)^2$.

Ms. Liston: Zoe, tell me about that. What was your thinking?

Zoe: Well, you need to get 16, and 2 times 8 is 16.

Sam: No, you get the 16 from 4 times 4 because the last number is a square number.

Zoe: Oh, yeah, you're right.

Ms. Liston: Great, what do others think? [*The class agrees with Sam and Zoe that $(x + 4)^2$ is correct.*] OK, let's move on to a new idea that will connect to what we've been doing. [*The lesson begins with familiarizing students with an arithmetic-geometry perspective.*]

Ms. Liston: Use your unit tiles to build a 3-by-5 rectangle [*see fig. 1.2*]. How many square units make up the area of this rectangle?

Zoe: 15.

Ms. Liston: How did you get 15?

Zoe: I just did 3 times 5.

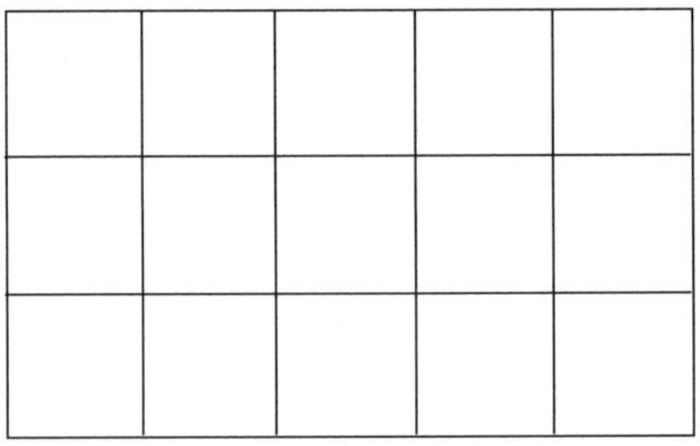

Fig. 1.2. A 3-by-5 rectangle (re-created from Vinogradova [2007, p. 404])

Ms. Liston: OK, very nice. Here's what I would like you to try to do: I would like to see if you can make a square out of this rectangle of 15 square units.

Sam: You can't.

Ms. Liston: Tell me, why do you say that?

Sam: Because 15 is not a square number.

Ms. Liston: But 15 is close to 16.

Sam: Yeah, it's one less.

Ms. Liston: Yes. Can everyone show me by rearranging the tiles how our rectangle of 15 unit squares could "almost" look like a square and record your ideas on paper?

[*The students draw and redraw and working together they come up with "squares" that have a missing piece. See fig. 1.3.*]

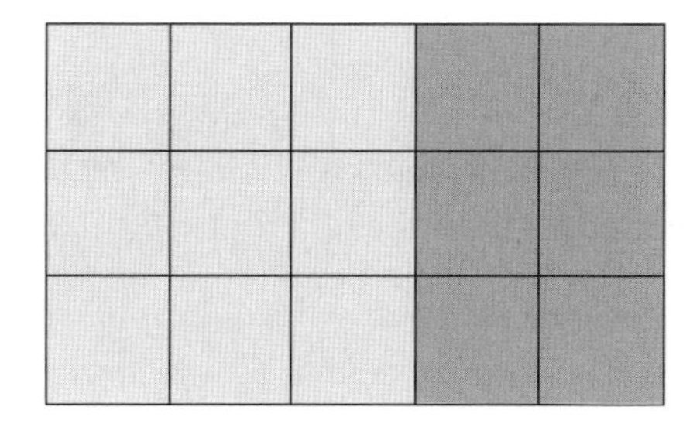

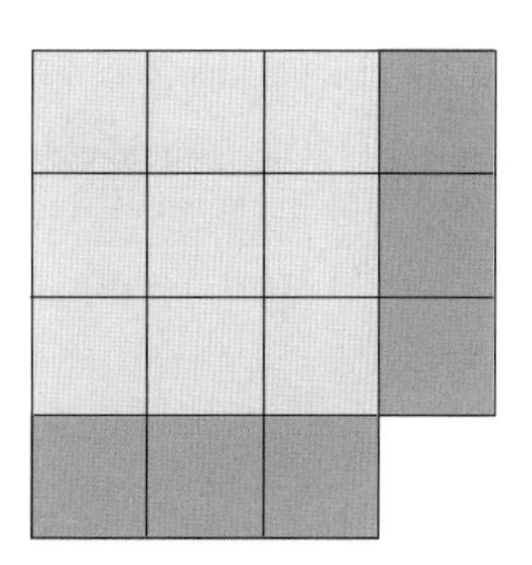

Fig. 1.3. Recomposing a rectangle into a "square" (re-created from Vinogradova [2007, p. 404])

Ms. Liston: Good job. One way to express our square with a missing piece is to write $15 = (3 + 1)^2 - 1$. Try making a square out of 22 square units. Remember, you might have some missing pieces. [*She lets the students work for a few minutes.*] Mateo, what did you get, and can you show us your drawing?

Mateo: I got $(4 + 1)^2 - 3$.

[*Students then practice the same process of completing the square using algebraic tiles.*]

Ms. Liston: So far we have been working with a rectangle with set dimensions and then making that into something that is "almost" a square. But what if we don't know the exact dimensions of the rectangle?

Mateo: We'll have to use variables.

Ms. Liston: How would you like to do that?

Mateo: We could use x and y.

Ms. Liston: OK. Will you come up to the board and draw your rectangle for the class?

[*Mateo comes to the board and draws a rectangle with y being longer than x.*]

Ms. Liston: How can we decide how long y is in comparison to x?

Mateo: It's longer.

Ms. Liston: How much longer?

Mateo: Let's make it two longer.

Ms. Liston: Can one of your group members tell us how to use algebra to write that relationship?

Jessica: In place of y you could put $x + 2$.

[*Students nod, so Ms. Liston continues.*]

Ms. Liston: Okay, would everyone use the tile blocks and build Mateo's rectangle so that it is x on one dimension and $x + 2$ on the other? [*See fig. 1.4.*]

[*Ms. Liston walks around and checks on the groups.*]

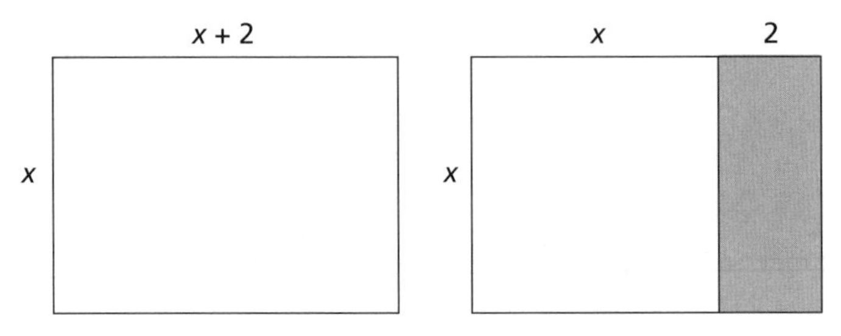

Fig. 1.4. Building an x by $x + 2$ rectangle
(re-created from Vinogradova [2007, p. 404])

Ms. Liston: Your tile rectangles look good. What's the area of your rectangle?

Zoe: Mine is $x(x + 2)$.

Sam: I got $x^2 + 2$.

Zoe: Well, you could do that, but you have to have $x^2 + 2x$.

Sam: That's what I meant to say.

Ms. Liston: Why are both answers, $x(x + 2)$ and $x^2 + 2x$ correct?

Sam: Because you distribute the x.

Ms. Liston: Great. What can we do to try to make a square out of our rectangle?

Sam: Well, you could move one of the x-rods like this. [*Shows moving x-rod as in fig. 1.5.*]

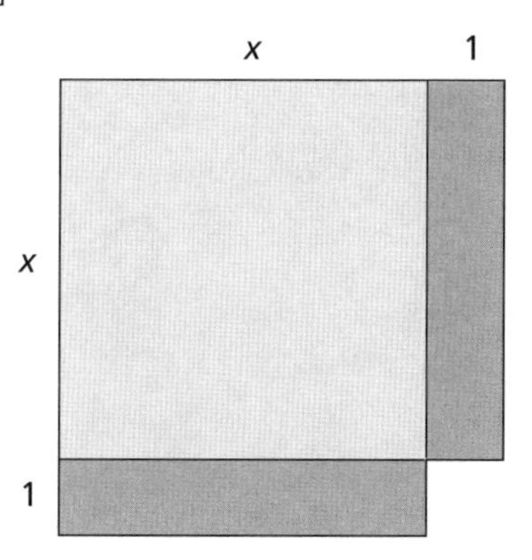

Fig. 1.5. Making a "square" out of the rectangle (re-created from Vinogradova [2007, p. 404])

Ms. Liston: Very nice. This is very close to a square now. How much is missing?

Sam: Only one unit-square.

Ms. Liston: I would like everyone to think about how to write the area of your rectangle, remembering that it is almost a square; it just has one unit-square missing.

Sam: The big square is $(x + 1)^2$. Then you have to subtract the 1, so $(x + 1)^2 - 1$.

Ms. Liston: That looks different from $x^2 + 2x$. Can you convince me that they are the same?

[*Students distribute $(x + 1)^2$, subtract 1, and find that the expressions are equivalent.*]

What is interesting about this introduction to completing the square is that with these examples, students begin to realize (by manipulation of the algebra tiles) that rectangles can be viewed as squares with missing pieces. In addition, the number of 1-by-x rods must always be divided into two parts. As the idea of CTS continues, students can be asked, what number of square units will complete the square? After the connection is made by students that "what's missing from the square" is what is needed to complete the square, the generalization in figure 1.6 can be made.

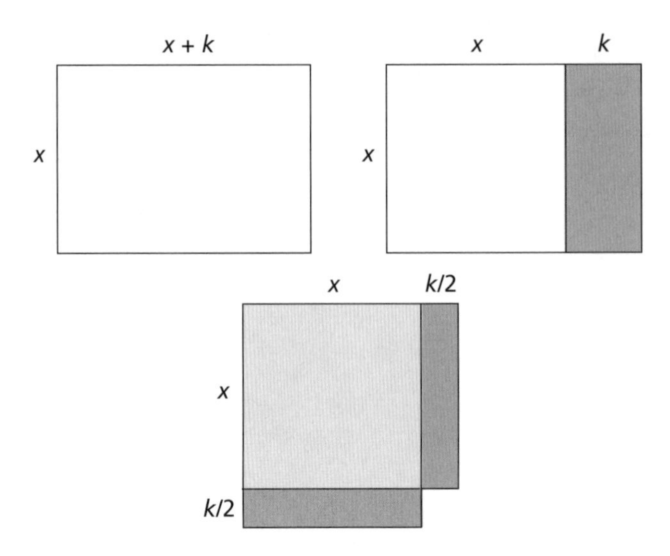

Fig. 1.6. Recomposing a rectangle into a "square" (re-created from Vinogradova [2007, p. 405])

These three vignettes show how students can be engaged in problems by making sense of them and persevere in solving them by using strategies that make sense to them and others. Students are able to develop and communicate their strategies and to modify their strategies if needed. In doing so, they construct a new understanding of the mathematics concepts involved, as well as strengthening their knowledge of related concepts.

Resources

This resource describes the teacher's role in supporting mathematical thinking and problem solving in the classroom by showing how teachers elicit student thinking, as well as promoting reflection and sense making.

- Rigelman, N. R. "Fostering Mathematical Thinking and Problem Solving: The Teacher's Role." *Teaching Children Mathematics* 13, no. 6 (2007): 308–14.

This resource presents a classroom vignette where a teacher poses a problem to students and highlights important aspects of her pedagogy that support English language learners (ELLs) in the problem-solving process.

- Wiest, L. R. "Problem Solving Support for English Language Learners." *Teaching Children Mathematics* 14 (April 2008): 479–84.

This resource describes pedagogical practices that can be used to support ELLs in understanding and succeeding in mathematics.

- Brown, C. L., J. A. Cady, and P. M. Taylor. "Problem Solving and the English Language Learner." *Mathematics Teaching in the Middle School* 14, no. 9 (2009): 532–39.

This resource describes a problem-centered curriculum for high school students that uses problem sets to engage students in problem solving. Students gain mathematical understanding of the concepts in the context of the problems.

- Campbell, W. E., J. C. Kemp, and J. H. Zia. "Bugs, Planes, and Ferris Wheels: A Problem-Centered Curriculum." *The Mathematics Teacher* 99, no. 6 (2006): 406–13.

This resource presents teaching through problem solving pedagogy that aims to engage students in problem solving as a tool to facilitate learning important mathematics subject matter and mathematical practices.

- Fi, C. D., and K. M. Degner. "Teaching through Problem Solving." *The Mathematics Teacher* 105, no. 6 (2012): 455–59.

References

Boaler, J., and C. Humphreys. *Connecting Mathematical Ideas: Middle School Video Cases to Support Teaching and Learning.* Portsmouth, N.H.: Heinemann, 2005.

Burns, M., and C. Humphreys. *A Collection of Math Lessons: From Grades 6–8.* New Rochelle, N.Y.: The Math Solution Publications, 1990.

Carpenter, T. P., E. Fennema, M. Franke, L. Levi, and S. Empson. *Children's Mathematics: Cognitively Guided Instruction.* Portsmouth, N.H.: Heinemann, 1999.

Carpenter, T. P., and R. Lehrer. "Teaching and Learning Mathematics with Understanding." In *Mathematics Classrooms that Promote Understanding,* edited by Elizabeth Fennema and Thomas A. Romberg, pp. 19–32. Mahwah, N.J.: Lawrence Erlbaum Associates, 1999.

Civil, M. "Building on Community Knowledge: An Avenue to Equity in Mathematics Education." In *Improving Access to Mathematics: Diversity and Equity in the Classroom,* edited by N. S. Nasir and P. Cobb, pp. 105–17. New York: Teachers College Press, 2007.

Common Core State Standards Initiative (CCSSI). *Common Core State Standards for Mathematics.* Washington, D.C.: National Governors Association Center for Best Practices and the Council of Chief State School Officers, 2010. http://www.corestandards.org.

Gutstein, E. *Reading and Writing the World with Mathematics: Toward a Pedagogy of Social Justice.* New York: Routledge, 2006.

Gutstein, E. "Connecting Community, Critical, and Classical Knowledge in Teaching Mathematics for Social Justice." In *International Perspectives on Social Justice in Mathematics Education,* edited by Barath Sriraman, pp. 109–18. Monograph 1 of *The Montana Enthusiast.* Montana: The University of Montana and the Montana Council of Teachers of Mathematics, 2007.

Hatano, Giyoo, and Kayoko Inagaki. "Sharing Cognition through Collective Comprehension Activity." In *Perspectives on Socially Shared Cognition,* edited by Lauren B. Resnick, John M. Levine, and Stephanie D. Teasley, pp. 331–48. Washington, D.C.: American Psychological Association, 1991. Cited in National Council of Teachers of Mathematics (NCTM). *Principles and Standards for School Mathematics.* Reston, Va: NCTM, 2000.

Hiebert, J., T. P. Carpenter, E. Fennema, K. C. Fuson, D. Wearne, H. Murray, A. Olivier, and P. Human. *Making Sense: Teaching and Learning Mathematics with Understanding.* Portsmouth, N.H.: Heinemann, 1997.

National Council of Teachers of Mathematics (NCTM). *Principles and Standards for School Mathematics.* Reston, Va: NCTM, 2000.

National Research Council (NRC). *Adding It Up: Helping Children Learn Mathematics.* Edited by J. Kilpatrick, J. Swafford, and B. Findell. Washington, D.C.: National Academies Press, 2001.

Peterson, P., Fennema, E., Carpenter, T. P., and Loef, M. "Teachers' Pedagogical Content Beliefs in Mathematics." *Cognition and Instruction* 6, no. 1 (1989): 1–40.

Vinogradova, N. "Solving Quadratic Equations by Completing Squares." *Mathematics Teaching in the Middle School* 12, no. 7 (2007): 403–5.

Reason Abstractly and Quantitatively

Practice 2: Reason abstractly and quantitatively

Mathematically proficient students make sense of quantities and their relationships in problem situations. They bring two complementary abilities to bear on problems involving quantitative relationships: the ability to *decontextualize*—to abstract a given situation and represent it symbolically and manipulate the representing symbols as if they have a life of their own, without necessarily attending to their referents—and the ability to *contextualize*, to pause as needed during the manipulation process in order to probe into the referents for the symbols involved. Quantitative reasoning entails habits of creating a coherent representation of the problem at hand; considering the units involved; attending to the meaning of quantities, not just how to compute them; and knowing and flexibly using different properties of operations and objects. (CCSSI 2010, p. 6)

Unpacking the Practice

Practice 2 highlights a general disposition and set of skills that we hope to see across all of mathematics. However, it has particularly strong ties to children's understanding of number and operations and algebra. While it is beyond the scope of this book to provide an in-depth analysis of connections to the National Council of Teachers of Mathematics (NCTM) Content Standards, after addressing the Process Standards we briefly highlight a few key connections to NCTM's Number and Operations Standard and Algebra Standard.

Problem Solving Standard

Practice 2 of the *Common Core State Standards for Mathematics* (CCSSM) begins with a focus on problem solving by highlighting how students should "make sense of quantities and their relationships *in problem situations*" (CCSSI 2010, p. 6; italics added), with a particular emphasis on the role of quantitative and abstract reasoning when solving problems. Although the practice is framed as a set of skills proficient students bring to bear on a problem, it is important to understand that abstract and quantitative reasoning develop *through* problem-solving opportunities. The first vignette in the examples section shows how problem-solving situations can be used to introduce students to or deepen their understanding of this practice. This connects to NCTM's Problem Solving Standard, which states that students should have opportunities to "build new mathematical knowledge through problem solving" (NCTM 2000, p. 52).

Another important aspect of NCTM's Problem Solving Standard is that students should "solve problems that arise in mathematics and *in other contexts*" (NCTM 2000, p. 52; italics added). Practice 2 emphasizes the relationship between contextualized problems and mathematical symbols, thus providing ample opportunity to connect mathematics to other contexts. This practice also highlights the importance of *Adding It Up*'s procedural fluency strand in problem solving. "*Procedural fluency* refers to knowledge of procedures, knowledge of when and how to use them appropriately, and skill in performing them flexibly, accurately, and efficiently" (NRC 2001, p. 121). While much of practice 2 focuses on contextualizing and decontextualizing (moving back and forth between the problem situation and its mathematical representation), this practice also states that students should "manipulate the representing symbols as if they have a life of their own, without necessarily attending to their referents" (CCSSI 2010, p. 6). As students develop a greater expertise and fluency with mathematical procedures, they can use the procedures as a tool without the need to constantly refer back to the problem situation. However, as this practice emphasizes, students need to check with reality frequently to be sure that their mathematical work remains connected to the problem context.

Representation Standard

Representation plays a central role in practice 2 and in the decontextualizing and contextualizing that the practice emphasizes in particular. This practice explicitly states that students should "abstract a given situation and *represent* it symbolically" (CCSSI 2010, p. 6; italics added) and that "quantitative reasoning entails habits of creating a coherent *representation* of the problem at hand...[and] attending to the meaning of quantities" (p. 6; italics added). Central to this practice is the use of the language of mathematics to represent a problem in a useful way, as well as the ability to make sense of symbolic representations of problems.

NCTM's (2000) *Principles and Standards for School Mathematics* states that "the term *representation* refers both to process and to product—in other words, to the act of capturing a mathematical concept or relationship in some form and to the form itself" (p. 67). This dual nature of representation can be seen most clearly in the first vignette in the examples section. On the one hand, the teacher emphasizes the products by creating various symbolic representations of the students' strategies and using these representations of the problem to move forward. On the other hand, the entire vignette is an example of the process "of capturing a mathematical concept or relation in some form" (NCTM 2000, p. 67) because the teacher continually asks the students to explain the relationship between the symbolic representations and the problem context.

The importance of mathematical representation is also captured by the strategic competence strand from *Adding It Up*, which states that "with a formulated problem in hand, the student's first step in solving it is to represent it mathematically in some fashion, whether numerically, symbolically, verbally, or graphically" (NRC 2001, p. 124). By emphasizing the importance of contextualizing and "attending to the meaning of quantities" (CCSSI 2010, p. 6), this practice draws a clear distinction between reasoning about relationships among quantities and what *Adding It Up* refers to as "number grabbing":

> Becoming strategically competent involves an avoidance of "number grabbing" methods (in which the student selects numbers and prepares to perform arithmetic operations on them) in favor of methods that generate problem models (in which the student constructs a mental model of the variables and relations described in the problem). (NRC 2001, p. 124)

Notice that the emphasis in this practice is on "making sense of *quantities* and their relationships in problem situations" (CCSSI 2010, p. 6; italics added), not blindly computing with numbers or algebraic symbols. It is important to understand that quantity should be seen as a measureable attribute of an object and thus different from numbers. Thompson (1993) describes quantitative reasoning in the following way:

> A prominent characteristic of reasoning quantitatively is that numbers and numeric relationships are of secondary importance, and do not enter in to the primary analysis of a situation. What is important is relationships among quantities. In that regard, quantitative reasoning bears a strong resemblance to the kind of reasoning customarily emphasized in algebra instruction. (p. 165)

More specifically, Thompson argues that one can reason quantitatively without assigning specific measurements (numbers) and the use of numbers does not necessarily imply that one is reasoning quantitatively. He gives the example of being able to determine whether you or someone else is taller without actually measuring either person's height.

Communication Standard

As written, practice 2 does not explicitly address mathematical communication. However, this practice is deeply connected to mathematical communication in two important ways. First, as illustrated in the elementary grades vignette in the examples section, engaging students in this practice provides ample opportunity for students to communicate about their mathematical thinking. When students must explain the relationship between the real-world context of a problem and their symbolic representation of that context, they must articulate their thinking to others and listen to and make sense of others' thinking and explanations. These ideas are reflected in the first three goals of NCTM's Communication Standard: Instructional programs… should enable all students to—

- organize and consolidate their mathematical thinking through communication;
- communicate their mathematical thinking coherently and clearly to peers, teachers, and others;

- analyze and evaluate the mathematical thinking and strategies of others. (NCTM 2000, p. 60)

Therefore, we can see how this practice is supportive of NCTM's Communication Standard. Having students participate in abstract and quantitative reasoning provides rich opportunities for them to engage in mathematical communication. In addition, having students engage in mathematical communication is a key means of providing students with opportunities to reason abstractly and quantitatively.

Second, we can use mathematics as a means of communication, or as the fourth goal in NCTM's Communication Standard states, instructional programs should enable students to "use the language of mathematics to express mathematical ideas precisely" (2000, p. 60). Reasoning abstractly and qualitatively provides students with the opportunity to do exactly this. By translating real-world problem contexts into mathematical symbols, students have an opportunity to engage in a particularly precise form of communication that is available through the language of mathematics.

Connections Standard

Practice 2 provides key opportunities for students to make mathematical connections. NCTM's Connections Standard emphasizes that students should be able to "recognize and use connections among mathematical ideas" (NCTM 2000, p. 64) as well as "recognize and apply mathematics in contexts outside of mathematics" (p. 64). This practice provides opportunities for both. The first vignette in the examples section highlights the possibility of making connections between mathematical ideas by examining multiple symbolic representations of the same mathematical pattern. This practice also emphasizes students using mathematical symbols to represent problem situations and the need for them to relate their mathematical work back to those situations, providing the opportunity to connect mathematics to real-world contexts. Both types of connections support the development of *Adding It Up*'s productive disposition strand: Making connections between mathematical ideas helps students learn that "mathematics is understandable, not arbitrary" (NRC 2001, p. 131) and connections to real-world situations can help students see that mathematics is "both useful and worthwhile" (p. 131).

Number and Operations Standard

Reasoning abstractly and quantitatively requires that students understand multiple "ways of representing numbers [and the] relationships among numbers" (NCTM 2000, p. 32) as well as "understand meanings of operations and how they relate to one another" (p. 32). This might include renaming numbers in productive or useful ways, such as renaming 37 from "3 tens and 7 ones" to "2 tens and 17 ones" when using the standard U.S. algorithm for subtraction or relying on properties such as the commutative and associative properties of addition to simplify computations when solving a problem. Such work would reflect the idea that students should be able to "manipulate the representing symbols as if they have a life of their own" (CCSSI 2010, p. 6) and would highlight the importance of students "knowing and flexibly using different properties of operations and objects" (p. 6).

Algebra Standard

Practice 2 is arguably most closely related to NCTM's Algebra Standard. The practice points to the importance of students' ability to "*decontextualize*—to abstract a given situation and represent it symbolically and manipulate the representing symbols as if they have a life of their own" (CCSSI 2010, p. 6) and to "*contextualize*, to pause as needed during the manipulation process in order to probe into the referents for the symbols involved" (p. 6). A prototypical example of this practice might involve translating a problem situation into algebraic notation (decontextualizing), manipulating those symbols to arrive at a solution, and then reinterpreting that answer in terms of the original problem situation (contextualizing). The NCTM Algebra Standard emphasizes similar forms of reasoning, stating that students must be able to "represent and analyze mathematical situations and structures using algebraic symbols" (NCTM 2000, p. 37) and "use mathematical models to represent and understand quantitative relationships" (p. 37).

Classroom Examples

The examples below can be adapted to work across a broad range of grades and highlight two different aspects of reasoning abstractly and quantitatively. The elementary grades vignette focuses heavily on helping students learn to move back and forth between problem context (the number of seats at a table) and mathematical representations of this situation (such as symbolic expressions and equations). The middle and high school vignette offers a series of sample tasks that can be used to help students develop their understanding of *rate of change* (how one quantity changes in relationship to another quantity) and provides a clear emphasis on reasoning about relationships among quantities as opposed to considering specific numerical values.

Elementary Grades Vignette: How Many Seats?

The example below can be adapted to a broad range of elementary and middle grades. Work with younger children could focus more heavily on the earlier part of the vignette in which children share and explain their strategies with concrete numbers; see, for instance, the Dot Square problem (NCTM Problem Solving Standard; NCTM 2000, p. 185), which gives an example of a similar problem for grades 3 to 5. Work with older children could focus more heavily on introducing algebraic notation and developing a variety of algebraic expressions to match this situation. The vignette draws heavily from work with middle school students found in Bishop, Otto, and Lubinski (2001), Lannin (2003), and Boaler and Humphreys (2005), all three of which provide excellent examples of similar activities, how they can be used in the classroom, and the strategies children develop for solving these types of problems.

Teacher: Look at this picture I have on the board [*fig. 2.1*]. We are going to figure out how many people can sit at a table made out of squares. In the picture, there are four people sitting at just one square table. Figure out how many people can sit in the next three pictures.

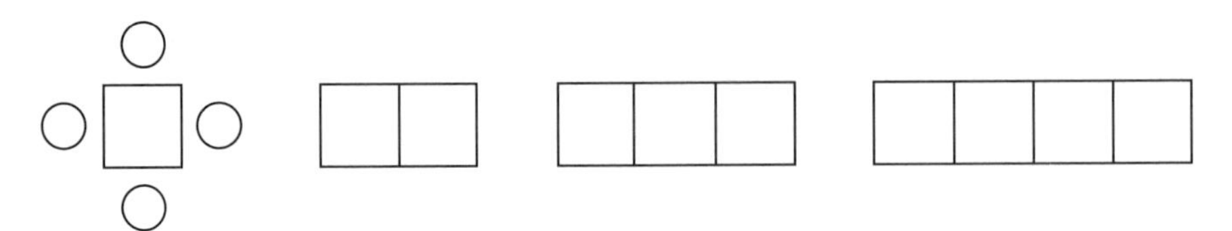

Fig. 2.1. Square table pattern

[*The students quickly complete this task, and the teacher has students explain their strategies for the fourth picture.*]

Teacher: Can someone come up and explain how they counted the number of seats for four squares?

Ashanti: Well this [*runs her finger along the top edge*] would be four chairs, and then the bottom would be another four, making eight. Then you would have these two [*points to the two ends*] makes ten.

Teacher: OK, did everyone see how she figured that out? I'm going to write that like this [*writes* 4 + 4 + 2]. Ashanti, do you think that shows how you counted up the seats, or should I write it a different way?

Ashanti: That's how I did it.

Teacher: Did anyone count the number of people in a different way?

Khalil: I counted three on each end, so that's six. Then I counted two for each table in the middle, so that's four more. Six and four is ten.

Teacher: I'm not sure I followed all that. Can you come up and show us what you mean by three on each end?

Khalil: I mean on this table [*points to the first square*] there are three seats, one, two, three [*touches the top, left, and bottom edges*]. And then the same thing for this table [*points to the last square*]. So that's six.

Teacher: OK, and then you said "two for each table in the middle"?

Khalil: Yeah, these two in the middle have two seats, one on the top and one on the bottom. So two twos is four more seats. So that's ten total.

Teacher: I see now. Can I write your strategy like this? [*Writes* 3 + 3 + 2 + 2.] Does that match your thinking?

Khalil: Yeah.

Teacher: OK, now I want you to imagine that you have ten square tables in a row. I want you to figure out how many people could sit there, but I want you to do it in three different ways. First, I want you to figure it out using

Ashanti's strategy. Then I want you to try Khalil's strategy. Then if you want you can come up with your own way.

[*The students work independently on this for a few minutes.*]

Teacher: OK, can someone besides Ashanti tell me how she would solve this problem?

Carlos: I can. First she would say there are ten seats all along the top. Then she would say there are ten more seats along the bottom. Then she would add two more seats for the ends. So that would be ten plus ten is twenty, plus two more is twenty-two.

Teacher: What do you think, Ashanti, does that fit your strategy from before?

Ashanti: Yeah, that's how I did it.

Teacher: OK, so if I write it the same way as before, so it kind of matches this [*points to* $4 + 4 + 2$ *written earlier*], but for this new problem, what would I write?

Juanita: It would be ten plus ten plus two.

Teacher: [*Writes* $10 + 10 + 2$ *next to* $4 + 4 + 2$.] OK, and what does this first ten mean?

Juanita: Those are all the seats along the top.

Teacher: OK, and how do you know to write a ten there? Did you count each seat by ones?

Juanita: Well, you could count them by ones, but you can also just know.

Teacher: How would you just know? Anyone can answer, how would you just know that there are ten seats along the top without counting by ones?

Jaylen: Every table has one seat on top, and there are ten tables, so you don't have to count, you just can figure out that it would be ten.

[*The conversation continues, and the teacher then facilitates a similar conversation about Khalil's strategy. The teacher then asks the class to repeat this activity for a row of one hundred tables, conducts another class discussion of the strategies, and then finally asks the whole class to consider if there were "n" tables pushed together.*]

Teacher: OK, so we have figured this out for ten tables and one hundred tables. What if we knew there were a lot of tables in a row, but we did not know exactly how many? When mathematicians have a problem like that, they use a letter to stand for the number of tables. So instead of saying there are ten tables or one hundred tables, they would say, "we don't know how many tables, so we will say there are n tables." And this n can be any number of tables. So think about how Ashanti would figure out how many people could sit down. Talk to your partners about how Ashanti could do this problem.

[*After a brief discussion in partners, the teacher calls the class back together.*]

Teacher: OK, any ideas? How could Ashanti use her strategy now? I know this is harder because we don't actually know how many tables there are. David, would you be willing to share what you and Rosa were talking about?

David: Well, we said it might be n plus n plus two.

Teacher: Where did you get that idea from?

David: Well, when it was four, it was four plus four plus two, and then with ten it was ten plus ten plus two, and the same with one hundred. So we just said "n" instead of the number.

Teacher: OK, so, I can write this $n + n + 2$, is that what you're saying?

David: Yeah, that's what we wrote down.

Teacher: OK, so, and anyone can answer this question, can someone explain what this means? What does this first n mean?

Rosa: Well, we were saying that the n is like how many seats there are on the top. Just like in the other problems.

Teacher: But how do you know how many there will be? How do you know it will be n?

Rosa: Because it's always the same as the number of tables.

[*The lesson continues with a transition to problems where the teacher tells the class the number of people and asks how many tables there must have been.*]

Practice 2 essentially describes a three-step process: (1) decontextualizing problems by representing a problem context using mathematical symbols; (2) manipulating symbols, such as performing calculations or solving an algebraic equation; and (3) contextualizing problems by periodically connecting the mathematical symbols back to the problem context. The vignette above focuses primarily on the first and third steps. The teacher solicits strategies and then demonstrates how they can be represented symbolically. However, she continually draws the students' attention to the meaning of the symbols in terms of the problem context (people sitting around a table). Thus the teacher is demonstrating how we can move back and forth between a problem context and a symbolic representation of that context.

Traditional mathematics instruction frequently involves teaching algebraic and symbolic rules first, without a meaningful problem context, and then later having students apply them to problem situations. This vignette shows how taking the opposite approach allows students to build meaning for mathematical symbols out of the sense making they have engaged in with the problem context (people sitting at a table), thus grounding the mathematical symbols in the students' reasoning instead of emphasizing an abstract set of rules and procedures to memorize.

Middle and High School Vignette: Rate of Change

Understanding rate of change is important for understanding functions and graphs and laying a foundation for calculus. An important aspect of understanding functional relationships involves coming to think of how one quantity (treated as the dependent variable) varies based on a relationship with another quantity (treated as the independent variable). Rate of change involves understanding how the dependent variable changes in response to changes in the independent variable. Is it increasing, constant, or decreasing, and is the rate of change steady, speeding up, or slowing down? A focus on rate of change can begin even before algebra, as can be seen in van Dyke and Tomback's (2005) article discussing how they collaborated to introduce algebra first through "qualitative graphs (graphs without scale), then quantitative graphs (graphs with scale)" (p. 237). The examples that follow are re-created from van Dyke and Tomback's work.

An early task can involve asking students to match graphs to a situation that is described in words, such as in figure 2.2. Initial problems such as these allow students to begin focusing on the relationship between quantities and when the dependent variable is increasing or decreasing as the independent variable increases. Later problems can provide a greater focus on the *intensity* of the rate of change, for instance considering when something is speeding up or slowing down as seen in figure 2.3.

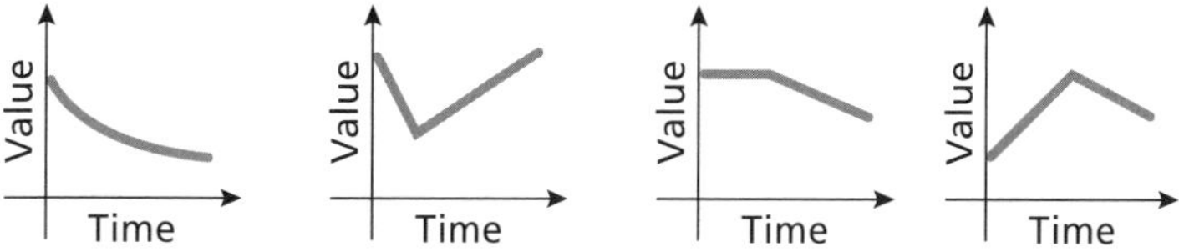

Fig. 2.2. Match a graph to a story
(adapted from van Dyke and Tomback [2005, pp. 237–38])

Janet, Gail, and Susan all walked away from the railroad station.
Janet walked at a steady pace, Gail speeded up as she walked,
and Susan slowed down.
Decide which graph pictures each girl's walk. Explain your reasoning.

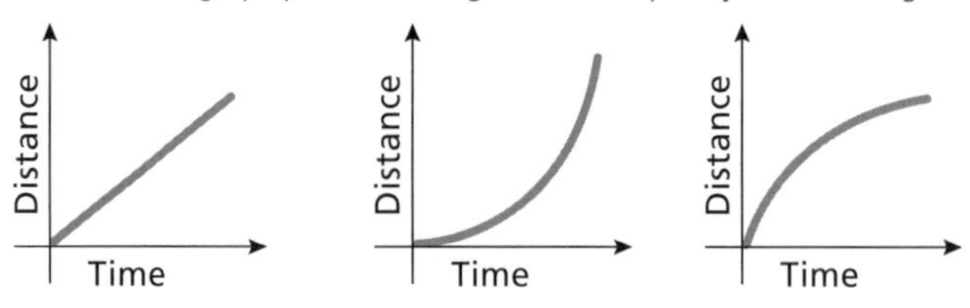

Fig. 2.3. Speeding up and slowing down
(re-created from van Dyke and Tomback [2005, p. 239])

Examples such as these are important for developing students' understanding of relationships between quantities and rate of change in particular. However, they also lay an important foundation for understanding calculus. Johnson (2012) provides an example of a student who developed such reasoning prior to taking calculus. Johnson found that rich understanding of rate of change involved the ability to systematically vary one variable, which acted as the independent variable (such as the side length of a square) and examine the change in the second variable (such as the area of the square). Importantly, as discussed in the examples above, the student could not only attend to the direction of change (was it increasing or decreasing) but to the intensity of this change (was it changing quickly or slowly, and was it a steady change or was it speeding up or slowing down).

One of the activities Johnson (2012) used was to create a square with the Geometer's Sketchpad software (Jackiw, 2001) and then allow the student to drag a corner of the square. The software would display the side length, the area, and the perimeter of the square. Johnson also later gave a set of two tables: one showing side length versus perimeter and another showing side length versus area. This activity allowed the student to explore how perimeter and area change in relationship to changes in side length. Two features of this problem that may help focus students' attention on the intensity of the rate of change are (1) students can examine when the area is growing faster than the perimeter and vice versa, and (2) students can investigate if the area changes at a steady rate or if it increases "faster and faster."

A common theme that cuts across these examples from van Dyke and Tomback (2005) and Johnson (2012) is the clear emphasis on the relationships among the various quantities *before* turning attention to the specific numerical values. This echoes the point raised above that quantity is not number (Thompson 1993); quantitative reasoning is fundamentally about considering the relationships among quantities within the problem context.

Resources

This resource provides a broad range of examples for incorporating algebraic reasoning into the elementary grades, including patterning activities and generalizing around simple story problems.

- Blanton, M. L., and J. J. Kaput. "Developing Elementary Teachers' Algebra Eyes and Ears." *Teaching Children Mathematics* 10, no. 2 (2003): 70–77.

These resources were drawn on heavily in the elementary-grades vignette above. They detail different forms of student thinking when making generalizations of patterns and the role of the teacher in accurately capturing students' thinking.

- Bishop, J. W., A. D. Otto, and C. A. Lubinski. "Promoting Algebraic Reasoning Using Students' Thinking." *Mathematics Teaching in the Middle School* 6, no. 9 (2001): 508–14.
- Lannin, J. K. "Developing Algebraic Reasoning through Generalization." *Mathematics Teaching in the Middle School* 8, no. 7 (2003): 342–48.

These resources connect to the secondary examples above in focusing on connecting qualitative graphs (graphs without scale) to real world contexts.

- Maus, J. "Every Story Tells a Picture." *Mathematics Teaching in the Middle School* 10, no. 8 (2005): 375–79.
- van Dyke, F., and J. Tomback. "Collaborating to Introduce Algebra." *Mathematics Teaching in the Middle School* 10, no. 5 (2005): 236–42.

This resource provides a clear emphasis on the relationship between algebraic symbols and problem contexts (the decontextualizing and contextualizing emphasized in practice 2). It includes problem contexts such as painting a room, identifying the number of dots in a star pattern, and summing consecutive numbers.

- Philipp, R. A., and B. P. Schappelle. "Algebra as Generalized Arithmetic: Starting with the Known for a Change." *The Mathematics Teacher* 92, no. 4 (1999): 310–16.

References

Bishop, J. W., A. D. Otto, and C. A. Lubinski. "Promoting Algebraic Reasoning Using Students' Thinking." *Mathematics Teaching in the Middle School* 6, no. 9 (2001): 508–14.

Boaler, J., and C. Humphreys. *Connecting Mathematical Ideas: Middle School Video Cases to Support Teaching and Learning.* Portsmouth, N.H.: Heinemann, 2005.

Common Core State Standards Initiative (CCSSI). *Common Core State Standards for Mathematics.* Washington, D.C.: National Governors Association Center for Best Practices and the Council of Chief State School Officers, 2010. http://www.corestandards.org.

Jackiw, N. The Geometer's Sketchpad (Version 4.0) [Computer Software]. Emmerville, Calif.: Key Curriculum Technologies, 2001.

Johnson, H. L. "Reasoning about Variation in the Intensity of Change in Covarying Quantities Involved in Rate of Change." *The Journal of Mathematical Behavior* 31 (2012): 313–30.

Lannin, J. K. "Developing Algebraic Reasoning through Generalization." *Mathematics Teaching in the Middle School* 8, no. 7 (2003): 342–48.

National Council of Teachers of Mathematics (NCTM). *Principles and Standards for School Mathematics.* Reston, Va: NCTM, 2000.

National Research Council (NRC). *Adding It Up: Helping Children Learn Mathematics.* Edited by J. Kilpatrick, J. Swafford, and B. Findell. Washington, D.C.: National Academies Press, 2001.

Thompson, P. W. "Quantitative Reasoning, Complexity, and Additive Structures." *Educational Studies in Mathematics* 25, no. 3 (1993): 165–208.

van Dyke, F., and J. Tomback. "Collaborating to Introduce Algebra." *Mathematics Teaching in the Middle School* 10, no. 5 (2005): 236–42.

Construct Viable Arguments and Critique the Reasoning of Others

Practice 3: Construct viable arguments and critique the reasoning of others

Mathematically proficient students understand and use stated assumptions, definitions, and previously established results in constructing arguments. They make conjectures and build a logical progression of statements to explore the truth of their conjectures. They are able to analyze situations by breaking them into cases, and can recognize and use counterexamples. They justify their conclusions, communicate them to others, and respond to the arguments of others. They reason inductively about data, making plausible arguments that take into account the context from which the data arose. Mathematically proficient students are also able to compare the effectiveness of two plausible arguments, distinguish correct logic or reasoning from that which is flawed, and—if there is a flaw in an argument—explain what it is. Elementary students can construct arguments using concrete referents such as objects, drawings, diagrams, and actions. Such arguments can make sense and be correct, even though they are not generalized or made formal until later grades. Later, students learn to determine domains to which an argument applies. Students at all grades can listen or read the arguments of others, decide whether they make sense, and ask useful questions to clarify or improve the arguments. (CCSSI 2010, p. 6)

Unpacking the Practice

Coming to know a discipline means understanding and engaging in arguments, which are connected or related sequences of statements to justify and refute claims in the discipline. As a discipline, mathematics places great emphasis on using valid, logical reasoning to establish whether a mathematical statement is true or false. The most rigorous, formal arguments are called *proofs*, whereas more informal arguments are called *justifications* and, in some cases, *explanations*. All proofs are justifications, some proofs also are explanations, but not all justifications and explanations are proofs. Learning how to argue whether an idea or claim is true or false in a mathematically valid way is an essential part of learning to do mathematics. Practice 3 builds on prior recommendations in the National Council of Teachers of Mathematics' Reasoning and Proof, Communication, and Representation Process Standards (NCTM 2000), as well as the mathematical proficiency strand of adaptive reasoning from *Adding It Up* (NRC 2001), to define goals for students' learning to (1) generate arguments and (2) evaluate arguments. This section aims to describe how these two aspects of practice 3 relate to, build upon, and distinguish themselves from existing recommendations for students' learning of argumentation and justification as discussed in *Principles and Standards for School Mathematics* (NCTM 2000) and *Adding It Up* (NRC 2001).

Reasoning and Proof Standard, Communication Standard, and *Adding It Up*

Across all levels—pre-K through grade 12—in its Reasoning and Proof Standard, NCTM calls for more opportunities for students to understand the purposes of reasoning and proving in mathematics, explore and make conjectures about relationships in patterns, provide reasoning to explain why conjectures are true or false, and use a variety of methods of mathematical proof. Like the Reasoning and Proof Standard, practice 3 highlights how students can build understanding of mathematical relationships by reasoning from examples, namely that students "can analyze situations by breaking them into cases, and can recognize and use counterexamples" (CCSSI 2010, p. 6). Although research on how students prove in mathematics classrooms indicates that they tend to overrely upon empirical, examples-based reasoning (Harel and Sowder 1998; Healy and Hoyles 2000; Hoyles and Küchemann 2002; Knuth, Choppin, and Bieda 2009), students intuitively reason "inductively about data" (CCSSI 2010, p. 6) and "teachers should guide them to use examples and counterexamples to test whether their generalizations are appropriate" (NCTM 2000, p. 122).

As students develop their ability to justify and prove, what *Adding It Up* (NRC 2001) refers to as a "manifestation" of adaptive reasoning, they move through three phases of reasoning: *empirical*, *preformal*, and *formal* (NCTM 2009). Empirical reasoning involves using examples to show that a claim is likely true. However, since confirming examples alone are insufficient for proof, students must learn to use definitions, assumptions, and previously proven theorems in constructing arguments that prove. Students reason at a preformal level when they generate "intuitive explanations and partial arguments that lend insight into what is happening" (NCTM 2009, p.10). Students who reason at a preformal level tend to use empirical evidence in a more generative way; they are able to describe some patterns, a general structure, or general features among the examples they have generated that can form a working conjecture or theory of the relationships between symbols or quantities. Although more challenging, all students should learn to engage in formal reasoning—generating logical arguments that represent valid mathematical proofs of a claim.

Practice 3 describes the kinds of activities students need to engage in, and the resources they must draw upon, to move from the empirical stage into preformal and more formal ways of constructing arguments, such as justifying and proving. Additional activities that provide opportunities for students to create mathematical arguments are (1) investigating and generalizing relationships from patterns; (2) developing definitions of mathematical objects through examination of physical representations and comparisons of the similarities across representative cases or sets of similar examples; (3) stating generalized relationships as conjectures in precise terms; (4) using counterexamples to disprove and refine conjectures; and (5) explaining why a conjecture is true by using definitions and mathematical properties. Unlike the *Principles and Standards* (NCTM 2000), the *Common Core State Standards for Mathematics* (CCSSM; CCSSI 2010) does not specify how these activities are a part of mathematics instruction at various grade bands. In the paragraphs that follow, we briefly review recommendations from the *Principles and Standards* for how these activities can be incorporated into instruction at all grade levels.

Students at the pre-K—grade-2 level can explore visual and numerical patterns, as well as begin to use basic mathematical properties, such as triangles have three sides, to justify claims (NCTM 2000, pp. 123–24). For example, students might use the fact that triangles have three sides to argue why a square, or other regular polygon, is not a triangle as well as why triangles have three, and only three, angles. Faulkner, Levi, and Carpenter (1999) illustrate that students at the K–2 level can produce mathematically valid arguments about relationships such as $a + b - b =$ a. However, at this stage, students' arguments may still consist primarily of reasoning from concrete objects or numerical examples. The transition into preformal reasoning begins in grades 3–5, when students become comfortable making conjectures and recognize that arguing a conjecture is true by simply providing a set of confirming examples is insufficient reasoning. Analyzing numeric and geometric patterns and stating generalizations about these patterns should be a regular part of students' mathematical work. At this level, students can develop more precise definitions and have access to a richer repertoire of properties and established truths to use as tools to construct justifications. An important role for teachers at this level is to scaffold students' use of definitions, properties, and assumptions by reminding students about what truths have been established and helping students keep track of conjectures that have been proved and those that remain to be proved (NCTM 2000, p. 190).

In grades 6–12, students can continue to investigate a variety of patterns and reason inductively about data; however, the justifications for conjectures should become progressively more rigorous. Increased rigor may be reflected in the level of precision used, the application of symbolic notation to indicate variables, and the format of the written argument. Although students at these grade levels should use examples to test conjectures, they should more regularly use formal reasoning to generate valid, deductive arguments. Throughout these grade levels, teachers play an important role in developing students' abilities to generate mathematically sophisticated arguments by continually pressing students to formulate conjectures and provide explanations as to why they are true (NCTM 2000, pp. 265–67). Even though the nature of mathematics at the secondary level is more abstract than concrete, students may exhibit reasoning across all three phases—empirical, preformal, and formal—when they learn new topics, such as geometry or calculus. Although all mathematical fields of study are based upon an interconnected set of definitions and properties, students may initially use more empirical ways of reasoning as they become familiar with new concepts. Encouraging and reminding students to draw upon definitions and properties can help them develop facility with understanding and applying abstract concepts, as well as expressing reasoning that treats a general case.

Evaluating Arguments

Practice 3 clearly states that students must understand how to evaluate arguments to attain mathematical proficiency. Efforts to understand others' arguments about a mathematical idea often support developing deeper conceptual understanding. Although mentions of the importance of learning how to evaluate arguments appear throughout *Principles and Standards for School Mathematics* (NCTM 2000), the Reasoning and Proof Standard does not unpack what learning to evaluate arguments entails and the teacher's role in supporting it as a mathematical practice. Competency in evaluating arguments requires that students (1) understand the implications of the conjecture or claim; (2) recognize the method of argumentation used and understand which methods are valid in mathematics (Weber 2010); and (3) infer *warrants*—the explanations why the reasons or statements given are adequate support for a conjecture or claim—or determine whether statements in the argument are based on valid principles (Weber 2010). Research suggests students may be able to determine when an argument is based on weak reasoning without being able to construct a mathematically valid argument on their own (Bieda and Lepak 2010) and can recognize that reasoning with examples is insufficient even though they may not know how to construct a valid proof of a statement (Weber 2010).

Engaging students in evaluating arguments requires that instruction either provide students with arguments to validate or, better yet, use arguments generated by other students in the class as objects to evaluate. The Reasoning and Proof Standard for grades 9–12 suggests ways to manage productive and considerate discussions when students evaluate others' arguments (NCTM 2000, pp. 342–46). In particular, the notion that ideas are being evaluated and not people is important to stress when students acknowledge shortcomings in others' reasoning. Additionally, the ways students choose to represent arguments can help or hinder other students in making sense of that reasoning. Building students' competencies to communicate their reasoning can work hand in hand with building students' competencies to evaluate others' reasoning. As stated in the Communication Standard, an important part of learning to communicate mathematics is to "use the language of mathematics to express mathematical ideas precisely" (NCTM 2000, p.

60). Describing ideas accurately can help students correctly interpret the meaning of terms used in another student's argument.

Classroom Examples

In this section, we provide vignettes—or episodes of mathematics instruction—to illustrate how practice 3 can be incorporated into elementary, middle, and high school mathematics instruction. Implementing practice 3 involves much more than simply asking students to explain why or to generate justifications. Each vignette for practice 3 illustrates that teaching to the practice requires thoughtful attention to the kinds of tasks that promote students' thinking about argumentation as well as discourse moves that lead to discussions where arguments are generated and evaluated. As you read through the vignettes, we encourage you to consider how your current teaching practice—the kinds of tasks you use during instruction, the ways you engage with students about their thinking, and how you facilitate classroom discussions about mathematics—compares and contrasts with the teaching representing in these vignettes and what changes you might make to strengthen students' proficiencies in generating and evaluating mathematical arguments.

Elementary Grades Vignette: The Identity Property of Multiplication

The curriculum of elementary grades mathematics is rich with foundational concepts, but often the need to teach basic skills overshadows the possibility of students learning about mathematical justification. When devices such as multiplication tables are used to teach number facts, students come to believe that textbooks, teachers, and adults in general determine whether something is true in mathematics and don't believe they can use their own reasoning skills (Flores 2002). In the elementary grades, students use empirically based, authority-based, and reasoning-based argumentation schemes (Flores 2002; Carpenter et al. 1999; Sowder and Harel 1998). Students can exhibit both empirically based and authority-based schemes when they are in the empirical stage; however, the authority-based scheme can be particularly problematic when students are encouraged to reason in more deductive and general ways. When students draw upon an authority-based scheme to justify their reasoning, they rely upon an external source to be an authority as to whether something in mathematics is true or not. As a result, they may come to believe that they do not have ownership of making sense and validating the truth of mathematical statements. It can be very frustrating for learners when they are asked to provide justification if they believe that it is sufficient justification to argue that a statement is true if their teacher, older sibling, or the textbook says it is true (Flores 2002). In the elementary grades, it is particularly important that students are encouraged to explain why basic facts and foundational properties are true so that they recognize they are capable of establishing a statement as a truth in mathematics.

The following vignette illustrates how elementary students can reason about the identity property of multiplication, facilitated by instruction that supports students' ownership of making sense of mathematics and prevents students from using authority-based reasoning. The multiplicative identity property says that the identity element, the number 1, can be multiplied with any number without changing the value of that number (i.e., $1 \times a = a$ for any value of a). This is a basic property of arithmetic, but understanding why it holds depends upon a solid understanding of what multiplication and division mean.

Ms. Gray: So, earlier this week we started exploring some new operations, which we called multiplication and division. Joseph, do you remember one of the multiplication problems we were working on yesterday?

Joseph: So, I remember doing two times five, and then two times ten. I remember that two times five is like five plus five, and you showed us how it was like taking five fingers of one hand and then adding those together with the five fingers of the other hand. But, we couldn't do that for two times ten, because we only have one group of ten between our two hands. So, you had us partner up with someone else, and we counted all the fingers on our hands to get, [*whispers "five, ten, fifteen, twenty"*] twenty.

Ms. Gray: Amira, I see your hand is raised. Do you want to add to what Joseph explained?

Amira: Yeah, so I remember that for ten times two, it was really like two times five times *another* two, because the first two times five is equal to ten.

Ms. Gray: Thank you Joseph and Amira. I'm so glad you remember this from yesterday. So, Joseph remembered that we did the problems two times five and then two times ten, and we were able to figure those out by doing five plus five and then ten plus ten. [*Writes* 2×5 *with* $5 + 5$ *beneath and* 2×10 *with* $10 + 10$ *beneath.*] Then, Amira said that two times ten is the same as two times five times another two because ten is equal to five times two. [*Writes* 2×5 *beneath* 10 *and then writes* $2 \times 5 \times 2 = 10 \times 2$.] Great. We discussed that the reason that two times five equals five plus five is because multiplication means that we add together two copies of five because we multiply five by two. So, if I told you to find three times five, what would you do? Raoul?

Raoul: You would add three copies of five: five plus five plus five.

Ms. Gray: OK, what if I asked you to find one times five. What would you do? Melissa?

Melissa: You add one copy of five, so five plus five.

Amira: Wait! Two times five is five plus five.

Joseph: I'm confused.

Ricky: Maybe one times five and two times five are the same. We just have to remember they are the same.

Ms. Gray: So, it would be pretty difficult to do mathematics if we had to remember that one times five and two times five are the same. Can we make sense of it? Does anyone not think that one times five is five plus five?

Maria: I was thinking that you only had one copy of five. Since there is only one copy, it is just five. Not five plus five, because that is two copies. So, I thought one times five is just five.

Ms. Gray: Does Maria's reasoning make sense? Luke, I see you nodding your head. Can you explain Maria's argument in your own words?

Luke: Well, we know that if you multiply two numbers, you add copies of one number. The number of copies you add is equal to the other number. So, one times five says that there is only one copy of five. So, the answer is just five.

The students in this vignette exhibit preformal reasoning in that they are able to reason about a specific example in a general way. In Luke's explanation, he states a definition of multiplication in a general way and then applies it to a specific case, 1×5. Maria also applied the definition of multiplication in her argument but did not initially state what multiplication means in a general way. The vignette also shows moves made by Ms. Gray that were critical to students engaging in preformal argumentation. One move happened at the beginning of the dialogue: Ms. Gray asked students to remember multiplication problems they had worked during the previous lesson and then reminded students of the definition of multiplication while summarizing and explaining students' responses ("because multiplication means that we add together two copies of five because we multiply five by two"). Although her mention of the definition was subtle in the interaction, it was an important move to remind students that they make sense of operations by referring to how they have defined the operation. The second critical move Ms. Gray made was to manage the discussion so that students revoiced other students' responses in their own words ("Can you explain Maria's argument in your own words?") and to encourage students to consider how they should be able to make sense of mathematics. In particular, Ms. Gray's response to Ricky served to remind students that they are capable of understanding how the system of mathematics works ("So, it would be pretty difficult to do mathematics if we had to remember that one times five and two times five are the same. Can we make sense of it?"). Reinforcing these conceptions insures against students developing persistent authority-based reasoning schemes.

Middle School Vignette: The Square Pattern Task

In the middle grades, students begin to use algebraic notation to express general relationships. Not only does algebraic notation support justifications that treat the general case, but it facilitates the development of precise definitions of concepts and quantities. Moreover, students can begin to understand how adaptations to patterns or functional situations result in corresponding changes to conjectures or generalizations and then modify prior arguments to account for such adaptations. To illustrate the rich nature of students' justifications in the middle grades, and the potential impact of justifying actions on their conceptual understanding of mathematics, we present a vignette inspired by a case of middle school students developing understanding of function and variable through pattern-based tasks as written in *Implementing Standards-Based Instruction: A Casebook for Professional Development* by Mary Kay Stein and colleagues (2009, see pp. 8–28).

Ms. Hutchins is a teacher at a math, science, and technology magnet middle school in Chicago. She has been teaching for ten years and has started to use patterns tasks to help students develop understanding of variables and functions. Inspired by the Standards for Mathematical Practice from the *Common Core State Standards for Mathematics* (CCSSI 2010), she has started to incorporate more tasks that ask students to generate arguments and use questioning patterns that press students to evaluate each other's mathematical reasoning. For this lesson, Ms. Hutchins has chosen to start with a discussion of the Square Pattern task.

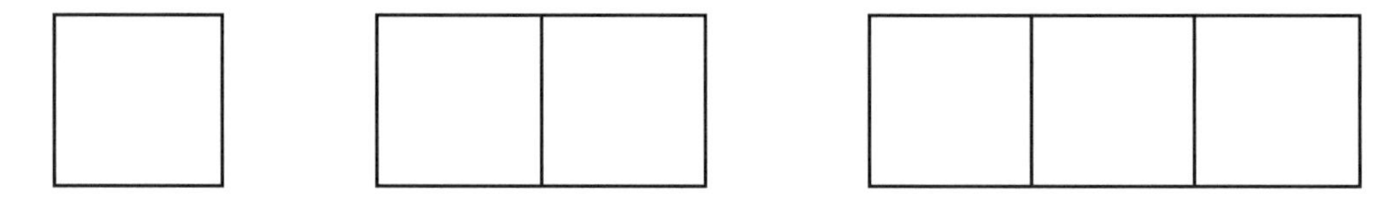

Fig. 3.1. The Square Pattern task

Ms. Hutchins: I've constructed a sequence of squares chained together [*see fig. 3.1*]. I'd like for you to think about how you would be able to find the perimeter of the chain, in units, if I were to chain 100 squares together. You can use the blocks at your table to construct chains, if you wish. You might also try constructing a table to see if you can notice a pattern.

[*The class works in small groups for ten minutes.*]

Ms. Hutchins: Andrea and Alex, I noticed your table has come up with a rule to find the perimeter of a 100-square chain. Would you tell the class your rule and why it makes sense to you?

Andrea: So, we figured out that each time a square is added, it adds 2 units to the perimeter. So, if the first step has a perimeter of 4 units, the second step has a perimeter of 4 + 2 or 6 units, and the third step has a perimeter of 4 + 2 + 2 or 8 units. The number of twos you add is always one less than the step number. So, for 100, the perimeter would be 4 + 2(99), which is 202 units. [*A student in the class, Markus, raises his hand*].

Ms. Hutchins: Markus, do you have a question for Andrea?

Markus: Yeah. Um, I'm kinda confused as to why she said that you only add 2 units to the perimeter when a square is added, because our group thought it was three. Like, you always have 1 unit on the end. Then, when you add another square, you add 3 more, but only the top and the bottom of the previous square count in the perimeter. So, you've got 1 plus 2 times the step number minus 1, and then 3 for the square added on the end. So, that's 4 plus 2 times the step number minus 1.

Andrea: But that's what I said, 4 plus 2 times 1 less than the step number. We've got the same rule.

Ms. Hutchins: You both may have the same rule, but you applied different reasoning to get the rule. Angelika, can you explain the differences between how Markus and Andrea justified their rule?

Angelika: Well, I think Markus was just seeing how the pattern was constructed differently. He saw 1 unit on one end and 3 units on the other end, and then the top and bottom of the middle units. Andrea's group saw it a little different. It was more like they saw a numeric pattern in how many twos get added on each time. They said they started with 4, then added 2 on each time.

Ms. Hutchins: OK, so can we figure out why that makes sense with the pattern and not just numerically?

Eyat: I think I see it. So, every time you add a square, you are really just adding a top and a bottom because there was already a right side to the square in the previous chain. The previous one is now in the inside of the chain but the new square adds a right side to the perimeter again. It's like it cancels out. So, all that is really new is the top and bottom sides of the new square, giving two more units.

Ms. Hutchins: Thanks, Eyat. Does that make sense, everyone? I'd like you to try to use the method used by Andrea's group with this next problem, because I'm interested to see if everyone can follow Eyat's explanation of Andrea's method. If you finish early, try to use the method Markus described with this new pattern. [*Ms. Hutchins posts a picture of a similar pattern to the Square Pattern task, except the figures are pentagons instead of squares (see fig. 3.2).*]

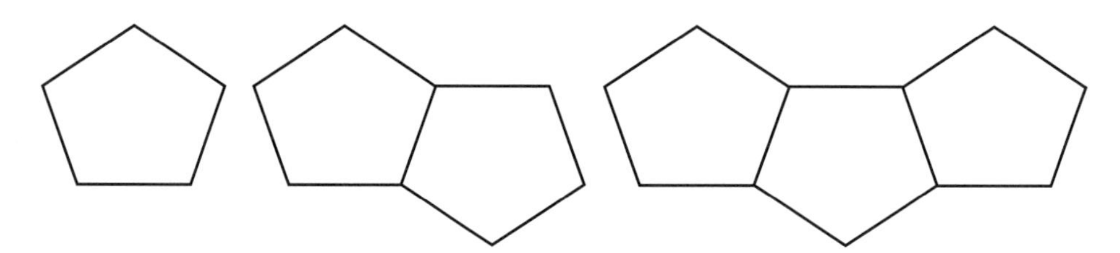

Fig. 3.2. Extending the Square Pattern task to pentagons

The vignette above depicts a situation where a class has been given a patterns task and asked to develop a rule to describe the relationship between the step number in the pattern and the perimeter of the figure in units. The students engage in small-group work to explore the pattern and generate a rule and then discuss their claims in whole-class discussion. Two aspects of the instruction depicted are critical to the engagement of all students in the process of creating and evaluating arguments. First, the task itself demands that students reason inductively from initial cases but then generalize claims to answer a question about what happens with the pattern when the step number is large. Second, Ms. Hutchins's orchestration of classroom discussion pushes students to consider others' claims and reasoning and to make choices that help students connect different solution methods and then apply their reasoning to a new task.

The Square Pattern task, although simple to describe and accessible to all learners, not only provides opportunities for rich connections to concepts such as variable and function but also supports multiple solution paths so that students must consider alternative ways to reason about the pattern. This is one key feature of tasks that support students in evaluating others' arguments—there are multiple approaches to describing a mathematical relationship that necessitate multiple (correct) ways of reasoning about the conjectured relationship. Without a task that allows for different approaches and is of high cognitive demand (Henningsen and Stein 1997), students cannot engage in the challenging thinking required to generate and evaluate arguments.

Ms. Hutchins's instructional moves—before students worked on the pattern task, during discussion about their strategies, and after strategies had been discussed—provided rich opportunities for students to engage in fundamental mathematical practices. In presenting the task, Ms. Hutchins provided concrete materials for students to use and encouraged students to use them. When asking students to share their rules, she also requested that students explain to the whole class why their rules made sense to them. In response to Andrea's comment about Markus's reasoning, she invited other students to make sense of Andrea's rule in terms of the visual pattern as a means of comparing and evaluating both methods. In giving a new task, she is assessing both students' understanding of the different ways to reason about the pattern as well as whether they can generalize their rule to a new pattern. These moves serve three main roles: (1) providing students access to reasoning about the task at both the empirical and preformal levels, (2) guiding discussion so that all students understand an argument before evaluating it, and (3) providing an extension to the original task that requires students to apply a generalization to a new problem.

High School Vignette: Heads and Tails

In grades 9–12, students should be expected to consistently engage in more formal ways of reasoning. Students' experiences with formal reasoning often occur in the context of high school geometry, where traditionally a more rigorous form of proof (i.e., the two-column proof) has been taught. It is becoming more prevalent for students to use a variety of representational forms for proof in high school geometry, including paragraph proofs and flowchart proofs. However, students' experiences with proof in high school algebra courses have been sparse in comparison. The significance of how practice 3 is positioned with respect to the rest of the Common Core Standards, namely that it is featured at the beginning of the document prior to discussion of the content standards, highlights how students' experiences with mathematical argumentation cannot be isolated to a particular grade level or course. New resources are available that offer specific suggestions for infusing argumentation throughout the high school mathematics curriculum including NCTM's *Focus on High School Mathematics: Reasoning and Sense-Making* (2009; see the resources at the end of the chapter).

The vignette we present illustrates how students can engage in generating and critiquing arguments in the context of high school probability and statistics. Students in the vignette have been working on combinations and permutations, and in this lesson they are determining how many possible permutations of heads and tails you get depending upon the number of tosses you make. (The vignette is adapted from a discussion of the relationship between outcomes of coin tosses and the binomial expansion at http://www.mathsisfun.com/pascals-triangle.html.) The class constructs table 3.1.

Table 3.1.
Relationship between number of tosses and possible outcomes

Number of Tosses	Possible Outcomes
1	H T
2	HH HT TH TT
3	HHH HHT HTH THH HTT THT TTH TTT
4	HHHH HHHT HHTH HTHH THHH HHTT HTHT HTTH THHT TTHH THTH TTTH TTHT THTT HTTT TTTT

Rico, a student in the class raises an interesting connection to Pascal's triangle and the coefficients of the binomial expansion that they worked on previously in an algebra course.

Rico: I think I see something interesting going on here. If you look, the number of outcomes for each condition is equal to those values for that triangle, I forget what it is called. You see, 1, 1; 1, 2, 1; 1, 3, 3, 1; 1, 4, 6, 4, 1

Mr. Miles: Oh, you mean Pascal's triangle, the one we looked at when we learned about the expansion of $(x + y)^n$?

Rico: I think so.

Tran: Yeah, I kind of remember that now.

Mr. Miles: That's an interesting observation. I'm not sure if I have seen that before! Why does it make sense that it would follow that pattern?

Rico: So, all you have to do is just add up the numbers to find the total possible outcomes. So, for three tosses it is 1 + 3 + 3 + 1 equals 8 possible outcomes.

Ally: Cool! Now we don't have to write them all out.

Mr. Miles: Hold up! We don't know if this is true yet. We need to justify why this would be true before we can just use it. Tran, do you have an explanation for why the values correspond to elements in Pascal's triangle?

Tran: Maybe it's related to that $(x + y)^n$ thing.

Mr. Miles: Oh, you mean the binomial expansion? [*Writes* $(x + y)^n$ *on the board with the phrase "binomial expansion" beneath.*]

Tran: What if we thought of x as heads and y as tails. Then, x plus y gives a heads and a tails. For $(x + y)^2$, it's $(x + y)(x + y)$, which is the same as $x^2 + xy + xy + y^2$. That's one outcome HH, two outcomes with one each HT and TH and one outcome TT.

Mr. Miles: That's interesting reasoning, Tran. Natalie, you have a question?

Natalie: OK, I'm really confused. What does Tran mean when he says x PLUS y "gives" a heads and a tails?

Alex: Yeah—I thought that Pascal's triangle gave the coefficients of the binomial expansion, not the actual terms.

Mr. Miles : These are excellent questions. Let's see if we can be more precise in our language. Natalie, you want to try to revise what Tran suggested heads and tails represent with the binomial expansion?

Natalie: You could think of the outcomes as equivalent to the number of x's and y's you have in the expansion, I guess. So, it's not that the outcomes with heads or tails equals x and y, but it is the number of x's and y's in the expansion.

In this vignette, Rico raises an interesting connection between the number of possible combinations of heads and tails in the coin tosses and the values in Pascal's triangle. The teacher, Mr. Miles, reminds the class briefly of their work with Pascal's triangle and its connection to the binomial expansion. When one student, Ally, assumes that the pattern Rico has found must always work, Mr. Miles steps in to caution the class about assuming something is true based on one case, a form of empirical reasoning. Mr. Miles then asks for justification, and Tran attempts to reason about why the outcomes correspond to values of coefficients in the binomial expansion. However, other students in the class question how Tran defined his terms in his argument, and Mr. Miles presses for alternative ways to define the related parts of the binomial expansion with the combinations of heads and tails.

Although in this example a student draws attention to an interesting relationship in the pattern of outcomes, teachers can also interject such conjectures for the class to explore. The vignette highlights how students can begin to ask other students to define their terms more precisely, but usually this norm is established with consistent input from the teacher using questions such as "How is Tran defining the relationship between the outcomes and x and y? Does this definition make sense?" The students do not construct a valid mathematical proof that the relationship Tran notices is true, but Mr. Miles's facilitation of the discussion about Tran's conjecture, in particular asking students to refine Tran's statements, supports the class in making sense of and evaluating Tran's argument.

With these vignettes, we hope we have provided situations that help unpack what it means to "teach to the practice." Although the Standards for Mathematical Practice in the *Common Core Standards for School Mathematics* (CCSSI 2010) are written as what students should understand and know how to do, it is the work of teaching to make sure that students have experiences where they build such knowledge and skills. The resource section below points to additional sources of information that can provide specific guidance on how to design tasks and orchestrate discussions that promote classroom argumentation.

Resources

Books

This book contains chapters written by many notable scholars in the area of reasoning and proving in school mathematics. Categorized by grade level, the chapters cover a range of issues from students' abilities to generate mathematical arguments and proofs to how teachers support collective argumentation in classrooms. This book is a comprehensive source of information on the most current theoretical and empirical work in the area of reasoning and proving in school mathematics.

- Stylianou, D. S., M. S. Blanton, and E. J. Knuth, eds. *Teaching and Learning Proof across the Grades.* Reston, Va: National Council of Teachers of Mathematics, 2009.

This book discusses the big ideas and essential understandings for proof and proving in grades 9 to 12. Like *Focus in High School Mathematics: Reasoning and Sense-Making* (NCTM 2009), this book provides an illustration of tasks and teaching techniques for incorporating proof and proving throughout the secondary mathematics curriculum. The book provides extensive discussion of the key mathematical issues involved in proving at the high school level.

- National Council of Teachers of Mathematics (NCTM). *Essential Understandings for Proof and Proving Grades 9–12.* Reston, Va: NCTM, 2012.

Journal Articles

The focus issues from three NCTM journals spanning K–12 mathematics (*Teaching Children Mathematics*, *Mathematics Teaching in the Middle School*, and *Mathematics Teacher*) provide both research-based and practice-based insights into teaching reasoning and mathematical argumentation throughout the school mathematics curriculum.

- *Teaching Children Mathematics.* Focus Issue "Reasoning and Sense Making," October 2011.
- *Mathematics Teaching in the Middle School.* Focus Issue "Fostering Mathematical Reasoning," February 2012.
- *Mathematics Teacher.* Focus Issue "Proof: Laying the Foundation," November 2009.

There is a wealth of research documenting students' difficulties with producing mathematical arguments that are deductive and based on reasoning about a general case. In this article, Stylianides and Stylianides provide insights on instructional sequences that can help students move past generating arguments with examples based on their work with such sequences in undergraduate mathematics content courses.

- Stylianides, G., and A. Stylianides. "Facilitating the Transition from Empirical Arguments to Proof." *Journal for Research in Mathematics Education* 40, no. 3 (2009): 314–52.

References

Bieda, K. N., and J. Lepak. "Students' Use of Givens When Proving: Context Matters." Paper presented at the National Council for Teachers of Mathematics Research Pre-Session, San Diego, Calif., April 2010.

Carpenter, T. P., E. Fennema, M. Franke, and S. Empson. *Children's Mathematical Thinking: Cognitively Guided Instruction.* Portsmouth, N.H.: Heinemann, 1999.

Common Core State Standards Initiative (CCSSI). *Common Core State Standards for Mathematics.* Washington, D.C.: National Governors Association Center for Best Practices and the Council of Chief State School Officers, 2010. http://www.corestandards.org.

Faulkner, K. P., L. Levi, and T. P. Carpenter. "Children's Understanding of Equality: A Foundation for Algebra." *Teaching Children Mathematics* 6, no. 4 (1999): 232–36.

Flores, A. "How Do Children Know that What They Learn in Mathematics Is True?" *Teaching Children Mathematics* 8, no. 5 (2002): 269–74.

Harel, G., and L. Sowder. "Students' Proof Schemes." *Research in Collegiate Mathematics Education* 3 (1998): 234–82.

Healy, L., and C. Hoyles. "A Study of Proof Conceptions in Algebra." *Journal for Research in Mathematics Education* 31 (2000): 396–428.

Henningsen, M., and M. K. Stein. "Mathematical Tasks and Student Cognition: Classroom-Based Factors that Support and Inhibit High-Level Mathematical Thinking and Reasoning." *Journal for Research in Mathematics Education* 28 (1997): 524–49.

Hoyles, C., and D. Küchemann. "Students' Understanding of Logical Implication." *Educational Studies in Mathematics* 51 (2002): 193–223.

Knuth, E., J. Choppin, and K. Bieda. "Middle School Students' Productions of Mathematical Justifications." In *Teaching and Learning Proof Across the Grades: A K–16 Perspective,* edited by M. Blanton, D. Stylianou, and E. Knuth. New York: Routledge, 2009.

National Council of Teachers of Mathematics (NCTM). *Principles and Standards for School Mathematics.* Reston, Va: NCTM, 2000.

National Council of Teachers of Mathematics (NCTM). *Focus in High School Mathematics: Reasoning and Sense Making.* Reston, Va: NCTM, 2009.

National Research Council (NRC). *Adding It Up: Helping Children Learn Mathematics.* Edited by J. Kilpatrick, J. Swafford, and B. Findell. Washington, D.C.: National Academies Press, 2001.

Sowder, L., and G. Harel. "Types of Students' Justifications." *Mathematics Teacher* 91, no. 8 (1998): 670–75.

Stein, M. K., M. S. Smith, M. A. Henningsen, and E. A. Silver. *Implementing Standards-Based Instruction: A Casebook for Professional Development.* New York: Teachers' College Press, 2009.

Weber, K. "Mathematics Majors' Perceptions of Conviction, Validity, and Proof." *Mathematical Thinking and Learning* 12, no. 4 (2010): 306–36.

Model with Mathematics

Practice 4: Model with mathematics

Mathematically proficient students can apply the mathematics they know to solve problems arising in everyday life, society, and the workplace. In early grades, this might be as simple as writing an addition equation to describe a situation. In middle grades, a student might apply proportional reasoning to plan a school event or analyze a problem in the community. By high school, a student might use geometry to solve a design problem or use a function to describe how one quantity of interest depends on another. Mathematically proficient students who can apply what they know are comfortable making assumptions and approximations to simplify a complicated situation, realizing that these may need revision later. They are able to identify important quantities in a practical situation and map their relationships using such tools as diagrams, two-way tables, graphs, flowcharts and formulas. They can analyze those relationships mathematically to draw conclusions. They routinely interpret their mathematical results in the context of the situation and reflect on whether the results make sense, possibly improving the model if it has not served its purpose. (CCSSI 2010, p. 7)

Unpacking the Practice

Mathematical modeling allows for a particular emphasis on the relationship between mathematics and the real world. Practice 4 builds on the National Council of Teachers of Mathematics' (NCTM 2000) Problem Solving and Connections Standards by emphasizing the role of mathematics in solving real world problems. Moreover, modeling provides students with numerous opportunities to engage in NCTM's Communication and Representation Standards as they develop, explain, and refine their mathematical models.

Problem Solving Standard

Mathematical modeling is fundamentally a problem-solving activity. As the introductory sentence of practice 4 tells us, modeling involves using mathematics "to solve problems arising in everyday life, society, and the workplace" (CCSSI 2010, p. 7). However, we should be careful not to interpret this too narrowly. Some curricula emphasize first learning mathematical concepts separate from real-world contexts and later "applying" these mathematical concepts to word problems. The practice of mathematical modeling is a much richer activity than this. While in some cases it may make pedagogical sense to address modeling after having learned about mathematical concepts, in many cases modeling can serve as motivation for and introduction to important mathematical concepts. As the NCTM Problem Solving Standard highlights, "good problems give students the chance to solidify and extend what they know and, when well chosen, *can stimulate mathematics learning*" (NCTM 2000, p. 52; italics added). Thus, mathematical modeling, and problem solving in general, can synthesize previous understanding and provide opportunities for new learning.

Modeling should go beyond word problems or application problems at the end of a unit or chapter. Students should regularly have opportunities to work with ill-defined problems. In discussing strategic competence, *Adding It Up* draws our attention to the fact that "outside of school they [students] encounter situations in which part of the difficulty is to figure out exactly what the problem is. Then they need to formulate the problem so that they can use mathematics to solve it" (NRC 2001, p. 124). Learning to formulate problems so that they can be tackled mathematically is an important part of modeling and addresses the practice's idea that "mathematically proficient students…are comfortable making assumptions and approximations to simplify a complicated situation."

In many respects, working with ill-defined problems is the key element of mathematical modeling, as it requires students to struggle to determine exactly how to use mathematics to better understand the problem at hand. Moreover, using ill-defined problems opens the door to the iterative nature of modeling. As stated in practice 4, students should be open to "possibly improving the model if it has not served its purpose." This is echoed in NCTM's Problem Solving Standard:

> Effective problem solvers constantly monitor and adjust what they are doing. They make sure they understand the problem.…They periodically take stock of their progress to see whether they seem to be on the right track. If they decide they are not making progress, they stop to consider alternatives and do not hesitate to take a completely different approach. (NCTM 2000, p. 54)

When modeling begins with ill-defined problems, students will have greater need to revisit their models and consider whether the model successfully captures the meaningful features of the problem before them. This provides opportunities for students not only to better understand the iterative nature of modeling but also to develop what *Adding It Up* calls a *productive disposition*: "the tendency to see sense in mathematics, to perceive it as both useful and worthwhile, to believe that steady effort in learning mathematics pays off, and to see oneself as an effective learner and doer of mathematics" (NRC 2001, p. 131).

Finally, in working through this iterative process, students can learn new mathematical knowledge or deepen existing understanding by refining mathematical concepts to fit the problem at hand. This can be seen in the vignettes in the "Classroom Examples" section, as students use and refine their understanding both of the real-world context and of the mathematical concepts involved through sharing different strategies and models and revising their work to better fit the situation.

Communication Standard

Although practice 4 does not explicitly call attention to the role of mathematical communication, the modeling practice and NCTM's Communication Standard can be mutually supportive. On the one hand, the practice of mathematical modeling provides opportunities for students to strengthen their mathematical communication, and on the other hand, students' understanding of modeling can be deepened by explicitly incorporating opportunities for engaging in communication. As stated above, modeling is fundamentally about using mathematics to solve problems in the real world. This provides a valuable opportunity for students to communicate their solutions and to open up these solutions to analysis by their peers. This communication can take a variety of forms. In early stages of the modeling process, students may communicate their ideas informally. However, as their peers and the teacher push for greater clarity and precision in the explanations, students can be guided to develop more formal explanations. This communication should draw on a variety of tools when appropriate, such as the "diagrams, two-way tables, graphs, flowcharts and formulas" that this mathematical practice highlights, as well as students' invented notations and representations when appropriate.

By communicating their modeling activities, students will not only improve their ability to communicate mathematical ideas; they will also deepen their understanding of the mathematical concepts they are studying. As their peers raise concerns, students may have to revise their mathematical models and their means of communication. This process helps students expand their mathematical repertoire while simultaneously reinforcing the idea that mathematical models are always provisional and that an important part of modeling involves "improving the model if it has not served its purpose" (CCSSI 2010, p. 7).

Representation Standard

Mathematical communication often goes hand in hand with representation. The need for students to communicate their thinking often provides opportunities for them to create, use, clarify, and refine ways of representing their mathematical ideas. In particular, the practice of mathematical modeling is related to NCTM's Representation Standard in two important ways. First, there is the role of representations, such as the "diagrams, two-way tables, graphs, flowcharts and formulas" discussed in this practice. Here, again, it is important to guard against a simplistic interpretation of the role of representations in mathematics in general and in modeling in particular. NCTM states that "the term *representation* refers both to process and to product—in other words, to the act of capturing a mathematical concept or relationship in some form and to the form itself" (NCTM 2000, p. 67). Therefore, when engaging students in mathematical modeling, representations should not be treated as prelearned tools that students have ready access to. In many cases, the act of modeling may create a need for a particular representation. In some cases these representations may be invented in the context of a modeling activity. As students invent ad hoc representations to serve the purposes of the modeling task, the teacher can also guide them toward using more standard representations.

Second, mathematical modeling itself can be understood as a form of representation. The NCTM Representation Standard states that "the term *mathematical model*...means a mathematical representation of the elements and relationships in an idealized version of a complex phenomenon" (NCTM 2000, p. 70). Through repeated modeling activities and explicit conversations about their strategies and thinking, students should come to see mathematics itself as a tool for representing real-world situations and problems in particular ways. Students should develop an awareness that their models highlight some aspects of the real-world situation while ignoring or approximating others as needed to solve any given problem.

Connections Standard

NCTM's Connections Standard states that "school mathematics experiences at all levels should include opportunities to learn about mathematics by working on problems arising in contexts outside of mathematics. These connections can be to other subject areas and disciplines as well as students' daily lives" (NCTM 2000, pp. 65–66). Clearly, mathematical modeling is a key means of accomplishing this goal, and as discussed earlier, seeing the value of mathematics in understanding real-world problems will develop students' productive disposition. As emphasized throughout this chapter, modeling does not involve superficial connections to real-world problems. Instead, mathematical modeling should involve real-world problems that are genuinely challenging for students on two levels. First, the students must be challenged to reformulate the real-world problem in mathematical terms. Second, the mathematics involved should not be trivial for the students. By keeping these two goals central to the practice of mathematical modeling, we strengthen students' understanding of the role of mathematics in the world and provide opportunities for them to develop rich understandings of mathematical concepts.

The role of connections in mathematical modeling makes this practice well suited to addressing issues of equity because it calls for drawing on problems from "everyday life, society, and the workplace." This allows for two important connections to the issue of equity in the classroom. First, modeling can be used as an opportunity to draw on genuine problems from students' everyday lives, thus making mathematics more relevant to students from all backgrounds. One powerful example of this can be found in the Funds of Knowledge work (e.g., González, Moll, and Amanti 2005; González et al. 2001). In the Funds of Knowledge work researchers and teachers learned about home and community resources and identified ways to make connections between these resources and the content being taught in the classroom. One example from the Funds of Knowledge work involving mathematics can be found in Civil and

Kahn's (2001) article in which a teacher used a gardening theme to explore area and perimeter in a way that was meaningful to the students. As Civil and Kahn point out, "Everyday contexts present children with authentic problems, such as how to make the [gardening] enclosure bigger, if we plan in-class activities that deliberately bring the mathematics to the foreground" (p. 404). While connecting mathematics to meaningful real-world or everyday contexts is valuable for all students, it can be of particular value for students of color or students from lower socioeconomic backgrounds whose perspectives have been traditionally left out of the mathematics curriculum.

The second way in which modeling can connect to issues of equity is through the use of mathematics to learn about, analyze, and challenge inequities in our world. Mathematics can be a powerful tool in learning to understand a variety of pressing social issues, such as national and global poverty, income and wealth inequality, global hunger, pollution, the fairness of different voting systems, and government spending. For instance, Tate (1994) gives an example of students who "formulated and proposed mathematically based economic incentives to get liquor stores to relocate away from the school" (p. 482). In another case, Gutiérrez (2009) describes students in an after-school program using mathematics to stop the closing of their local school. The students did such things as analyze test data, conduct a survey of their peers, and calculate average walking times to the new school. In other words, they identified how they could use mathematics to engage with a real-world issue they faced. The books *Rethinking Mathematics* by Gutstein and Peterson (2005) and *Math That Matters* by Stocker (2008) provide numerous examples of engaging students in using mathematics to understand social and political issues in students' local and global lives.

Classroom Examples

The three vignettes below highlight different aspects of mathematical modeling. The elementary grades example illustrates how modeling can be seen in student-invented strategies and how this can be built on by teachers to introduce more formal mathematical concepts. The middle school example is a problem designed specifically to engage students in modeling and is therefore particularly valuable for highlighting the iterative nature of modeling as well as the way in which mathematical concepts can be developed and solidified through modeling activities. Finally, while the high school example is more directed and of a more limited scope than the middle school problem, it highlights one way that teachers might intervene to guide modeling activities toward specific mathematical goals while also illustrating the value of having to make sense of formulas in real-world contexts.

Elementary Grades Vignette: Cognitively Guided Instruction

One particularly powerful example of the role modeling can play in elementary students' learning of mathematics can be found in cognitively guided instruction (CGI), which is introduced in an accessible manner in *Children's Mathematics: Cognitively Guided Instruction* by Carpenter and colleagues (1999). CGI examines the different types of problem situations that can be connected to the four basic operations (addition, subtraction, multiplication, and division), the progression of strategies that children naturally use in response to these problems (without direct instruction from the teacher), and how children's thinking about multidigit numbers develops. It is important to keep in mind that although this research was done with children in kindergarten through grade 2, the content of multiplication, division, and place value is found in the upper elementary grades as well. This vignette is adapted from ideas detailed in *Children's Mathematics.*

A teacher writes the following problem on the board and reads it out loud to the class: "I have 3 toys. My friend gives me some more. Now I have 7 toys. How many toys did my friend give me?" The teacher has never explicitly taught the students in her classroom how to solve this problem, nor has she spent much time drilling math facts. Instead, she allows the students to come up with and explain their own strategies. The teacher circulates around the room while the children work and asks several children to share their strategies. Three typical strategies indicate a range of mathematical abstractness:

- **Direct Modeling:** One student counts three blocks, placing them in front of her one at a time. She then adds additional blocks, keeping this new pile separated from the original pile. As she adds the new blocks she continues her counting, four, five, six, seven. She then stops and counts the new blocks she has added and announces that the answer is "Four. Your friend gave you four more toys."
- **Counting:** A second student says "three [pause], four, five, six, seven." The child extends one finger with each number in the counting sequence, starting with four. The child then looks at his fingers and announces that the answer is four.
- **Derived Facts:** A third student says "I know that three and three is six, and you would need one more to get to seven, so the answer is four."

After each child shares his or her thinking, the teacher asks if other students used similar strategies. The teacher then writes the following number sentence on the board: $3 + \square = 7$ and asks the class if this number sentence matches the strategies that the children shared.

Notice that although many adults may think of this problem as a "subtraction" problem, all three of the strategies presented are more closely related to the addition operation, as all three students figured out how much they had to add to the starting amount of three to get to the end amount of seven. This demonstrates children's natural inclination to model real-world problems, in that all three children developed a model that fit the action in the story problem. The teacher then built on the children's thinking by introducing a way of symbolizing the problem in the form of an open number sentence. This example illustrates how children can develop important mathematical concepts through modeling contextualized problems, can use this as an opportunity to communicate their ideas to their peers, and be introduced to a more mathematically formal representation of the concepts involved. This example also highlights the important role of the teacher in introducing conventional mathematical notation and representations as a way to build on students' informal mathematical models.

Middle School Vignette: The Quilt Problem

Lesh and Harel (2003) describe a series of modeling problems used with middle school students, one of which is the Quilt Problem, in which students explored how to design a real-sized quilt from a photograph. By the time they began this problem, the middle school students had already engaged in two previous modeling activities, so they had developed some comfort with the open-ended nature of the problems. While the vignette below is modified from the problem Lesh and Harel describe, the ideas and the nature of the students' reasoning is drawn from their work. This vignette highlights that many modeling activities require students to "routinely interpret their mathematical results in the context of the situation and reflect on whether the results make sense, possibly improving the model if it has not served its purpose" (CCSSI 2010, p. 7) as well as how students can develop conceptual understanding through mathematical modeling.

Teacher: Today we are going to work on coming up with a design for a quilt square. Sometimes quilters see a photograph of a quilt they would like to create, but they have trouble trying to re-create it themselves, so we are going to help them out. I'm handing out a picture of a quilt square [*see fig. 4.1*]. The actual size should be 12 inches by 12 inches. You have to write a letter explaining how to make templates of all of the pieces in this quilt square that are the correct size and shape, and you have to show how to follow your directions by making an example quilt square using construction paper.

Fig. 4.1. Quilt square design

[*A group of three students begins work by first attempting to find a scale factor that will enlarge the picture to its full size.*]

Andy: OK, I measured the square all the way across and it was 3 $^3/_8$ inches across. It's supposed to be 12 inches across. So [*typing on calculator*], $^3/_8$ is .375, so 12 divided by 3.375 is 3.555....That can't be right!

Larissa: We probably just need to measure more closely. Measure to the smallest mark, to the sixteenths.

[*After several attempts, the group continues to run into these messy numbers and instead switches their focus to the individual pieces in the quilt.*]

Parker: This isn't working. We keep getting the wrong numbers. Let's just say it's 3.5 and try that out. I'll measure the square, you measure the diamond, and you measure the triangles, and let's try seeing how they turn out after we multiply them by 3.5.

[*The group follows Parker's suggestion, measuring each of the shapes individually, multiplying the dimensions by 3.5, and then cutting them out of construction paper to piece together to make a larger quilt. At this point the students are not coordinating between the different pieces. For instance, they do not recognize that the short sides of the parallelogram, which they call a diamond, must be the same length as the sides of the square. They also do not have a means for finding the correct angles in the triangle and the parallelogram.*]

Parker: Ugh! Our pieces aren't fitting together right. See, this part on the diamond is too long, it doesn't match up with the edge of the square.

Larissa: Yeah, and there are gaps between the diamond and the triangle.

Andy: OK, well let's try it again, but measure as carefully as you can to get it right.

[*The students work for a little while trying to measure more carefully.*]

Andy: It's still not working! No matter how well we measure it doesn't fit right.

Larissa: Yeah, the side of the diamond has to fit perfectly with this square. These two parts have to be the same length.

Parker: OK, well let's change the diamond so it has shorter sides. Like we'll just measure the square and make the diamond the same size as that.

[*The group continues to work in this vein by using the small square as the basic shape and identifying relationships between the small square and the other two shapes. For instance, they realize that the triangles can be formed by cutting a square in half diagonally so the long side of the triangle is double the length of the side of the square. However, they continue to have difficulty in creating the proper angles for the parallelograms.*]

Larissa: These diamonds keep turning out wrong. There's always a little overlap or a gap when we try to fit them in.

Andy: What if we put the squares and triangles together and *then* make the diamonds so they fit in the empty spaces?

As the students continue to work, they begin identifying other ways in which the shapes in the quilt square are interrelated. For instance, in creating the parallelograms to fill in the empty spaces, they begin focusing on diagonal lines connecting the corner squares. Ultimately, focusing on how to construct the parallelograms leads the students to identify several important relationships based on the symmetries of the square. This also allows them to transition from thinking about building the whole out of the parts (starting with the small square) to thinking about how to create the smaller pieces so that they correctly fit into the whole. Their final directions begin by folding a 12″ by 12″ square along its diagonals and then horizontally and vertically in half. They then measure out a 3″ by 3″ square from each corner and use these to guide the construction of the rest of the quilt square (see fig. 4.2).

Notice that this problem highlights not only the iterative nature of mathematical modeling, in that the students continually revisited their method for scaling up the quilt square, but also the mathematical learning that can occur through modeling. When the students first approached the problem, they failed to coordinate between the various pieces that make up the whole. By the end of the lesson, the students had come to a strategy that manages to overcome the inaccuracy and messiness of the individual measurements while maintaining the appropriate relationship between the parts and the whole of the quilt square.

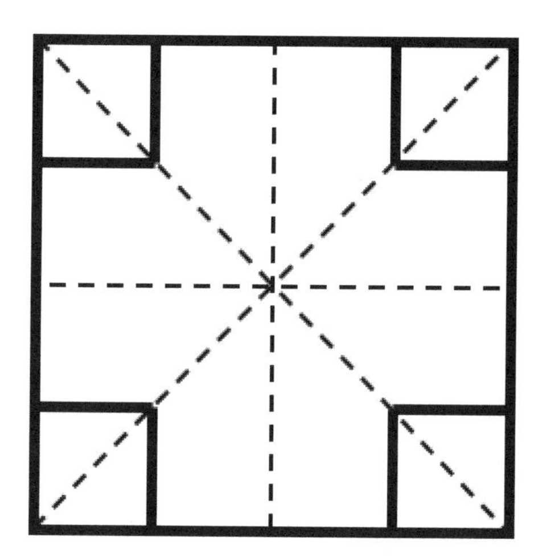

Fig. 4.2. Quilt square diagonals and corner squares

High School Vignette: South Central Los Angeles

Brantlinger (2005) describes his adaptation of a lesson developed by Gutstein (2005) in which high school students use mathematics to analyze the community resources available in South Central Los Angeles in 1992 when the Rodney King riots occurred. The following vignette is one possible version of how a class might explore these issues, building on ideas from Bratlinger and Gutstein and the experiences of Mathew Felton, one of the authors of this book, working with this content with preservice teachers.

After a brief discussion about what the students know about Rodney King and the 1992 riots, the teacher describes the task to the students (the following link provides some basic background information on the Los Angeles riots surrounding the Rodney King verdict: http://learning.blogs.nytimes.com/2012/05/01/may-1-1992-victim-rodney-kings-asks-can-we-all-get-along/).

Teacher: We are going to explore how many movie theaters, liquor stores, and community centers there were within a three-mile radius of South Central Los Angeles. One way to think about this question is by figuring out how many blocks, on average, someone would have to walk to reach one of these places in the neighborhood. We will be estimating, but to be sure that our estimates are not wild guesses, you should use mathematics to figure out how many city blocks there are in a three-mile radius.

[*The students begin by thinking primarily in linear terms. The teacher asks the groups to share their progress thus far with the class.*]

Kevin: The diameter of the circle is 6 miles, and there are 16 blocks per mile, so that's 96 city blocks.

Alison: It needs to be double that because you have to fill in both sides of the diameter.

Teacher: OK, so we have two different ways of looking at this problem. A helpful thing to do, especially with real-world problems like this, is to connect it back to the original situation and see how our strategy matches up. So can you show me where on the map these 96 or 192 blocks would be?

As the students begin counting, they quickly realize their estimates are far too low. However, the teacher's intervention has drawn the students' attention to the fact that they must go beyond measuring individual lengths to solve this problem. As the groups continue to work, they settle on using the formula $A = p(48)^2$ to find the number of city blocks that will fit in the three-mile radius (because 16 blocks fit in one mile, and thus 48 blocks fit in a three-mile radius).

The lesson continues with students making estimates about how many liquor stores, movie theaters, and community centers they expect to find in this three-mile radius by thinking about how far they would expect to walk to reach these kinds of buildings in their own neighborhoods and making comparisons to data the teacher has gathered about nearby neighborhoods. Eventually, the teacher shares the data he heard in a news story about this topic: there were no movie theaters or community centers but 640 liquor stores in this area at the time of the riots—values which are significantly different than those in their own neighborhoods. The students then find that there was approximately one liquor store for every 12 blocks and compare this to how far they expected to have to walk to various buildings in their own communities.

This vignette highlights a number of important aspects of mathematical modeling. It illustrates an example in which students are able to use mathematics to develop a preliminary understanding of a social issue. This is a complex issue, as other factors, such as population density, should be taken into account, and the reasons why this distribution of resources exists is also complex. However, it provides an example of using mathematics to begin to understand how the resources available may differ dramatically across different communities. Ideally, such an investigation would be deepened by connections to other disciplines and course work the students are involved in. Engaging students in using mathematics to address these issues can be challenging and Bratlinger (2005) describes two major difficulties he experienced when he taught a version of the South Central lesson. First, he became concerned that the lesson might reinforce the idea that there is something wrong with the people who live in South Central Los Angeles, and in future lessons he plans to bring in additional resources to focus on the underlying causes of the riots, such as economic disinvestment in the area. Second, he "shied away from opening up the political whole-class conversation that might have happened at the end of the lesson," (Bratlinger 2005, p. 100), instead talking more about his own beliefs and moving on to individual reflections. Bratlinger also notes that some of the students actively resisted the inclusion of political topics in the mathematics classroom.

This vignette also demonstrates that in many cases learning decontextualized formulas leaves students unprepared for using mathematics to make sense of real-world situations, as seen in the students' initial difficulty in determining the number of city blocks in the three-mile radius. This highlights the importance of practice 4: modeling real-world situations provides opportunities for students to determine what mathematics to draw on and how it is relevant to the problem at hand. Finally, this vignette also illustrates the role a teacher can play in helping students learn to model. In this case the teacher reminds the class that they need to connect their mathematics back to the real-world context, and in doing so, the teacher helps the students identify an important flaw in their models without taking over the problem-solving process.

Resources

This resource provides a rich example of how mathematical modeling can be used with elementary children. The authors provide an example of modeling where cyclones hit and analyzes the children's learning as well as the role of the teacher.

- English L., J. Fox, and J. Watters. "Problem Posing and Solving with Mathematical Modeling." *Teaching Children Mathematics* 12, no. 3 (2005): 156–63.

These three resources do not focus on modeling explicitly, but they emphasize the connection between mathematics and students' lives. The first provides an example of connecting area and perimeter to gardening, the second describes students taking pictures in their community and then writing their own mathematical problems, and the last emphasizes students posing their own mathematical problems related to community businesses.

- Civil, M., and L. Kahn. "Mathematics Instruction Developed from a Garden Theme." *Teaching Children Mathematics* 7, no. 7 (2001): 400–405.
- Leonard, J., and S. Guha. "Creating Cultural Relevance in Teaching and Learning Mathematics." *Teaching Children Mathematics* 9, no. 2 (2002): 114–18.
- Simic-Muller, K., E. E. Turner, and M. C. Varley. "Math Club Problem Posing." *Teaching Children Mathematics* 16, no. 4 (2009): 206–12.

The following resource provides a rich example of modeling with middle school students drawn from the same body of research as the Quilt Problem vignette above. In this article, the students work on the Big Foot activity, in which students must develop a procedure for predicting people's height based on their footprint.

- DiMatteo, R. W. "A Model Approach to Problem Solving." *Mathematics Teaching in the Middle School* 16, no. 3 (2010): 132–35.

This resource provides an example of preservice middle school teachers using modeling to provide arguments in favor of convoying ships during World War I.

- Mathews, S. M. "Mathematical Modeling: Convoying Merchant Ships." *Mathematics Teaching in the Middle School* 9, no. 7 (2004): 382–91.

There are a larger number of modeling resources for high school grades. The October 2007 issue (vol. 101, no. 3) of *Mathematics Teacher* includes several articles related to mathematical modeling. The three below are of particular relevance. The first shows the role technology can play in simulating animal populations over time. The second article emphasizes the potential for using statistical analysis to predict election outcomes. The third article provides examples of modeling simple physics experiments, such as modeling the height of a soccer ball dropped above a motion detector.

- Sinn, R. "Ecosystem Simulations and Chaos on the Graphing Calculator." *Mathematics Teacher* 101, no. 3 (2007): 167–75.
- Lamb, J. H. "Who Will Win? Predicting the Presidential Election Using Linear Regression." *Mathematics Teacher* 101, no. 3 (2007): 185–92.
- Hubbard, M. "Creating and Exploring Simple Models." *Mathematics Teacher* 101, no. 3 (2007): 193–99.

References

Brantlinger, A. "The Geometry of Inequality." In *Rethinking Mathematics: Teaching Social Justice by the Numbers*, edited by E. Gutstein and B. Peterson, pp. 97–100. Milwaukee, Wis.: Rethinking Schools, 2005.

Carpenter, T. P., E. Fennema, M. Franke, L. Levi, and S. Empson. *Children's Mathematics: Cognitively Guided Instruction.* Portsmouth, N.H.: Heinemann, 1999.

Civil, M., and L. Kahn. "Mathematics Instruction Developed from a Garden Theme." *Teaching Children Mathematics* 7, no. 7 (2001): 400–405.

Common Core State Standards Initiative (CCSSI). *Common Core State Standards for Mathematics.* Washington, D.C.: National Governors Association Center for Best Practices and the Council of Chief State School Officers, 2010. http://www.corestandards.org.

González, N., R. Andrade, M. Civil, and L. Moll. "Bridging Funds of Distributed Knowledge: Creating Zones of Practices in Mathematics." *Journal of Education for Students Placed at Risk* 6, no. 1 & 2 (2001): 115–32.

González, N., L. Moll, and C. Amanti. *Funds of Knowledge: Theorizing Practices in Households, Communities, and Classrooms.* Mahway, N.J.: Erlbaum, 2005.

Gutiérrez, M. V. "'I Thought This U.S. Place Was Supposed to Be about Freedom': Young Latinas Engage in Mathematics and Social Change to Save Their School." *Rethinking Schools* 24, no. 2 (2009): 36–39.

Gutstein, E. "South Central Los Angeles: Ratios and Density in Urban Areas." In *Rethinking Mathematics: Teaching Social Justice by the Numbers*, edited by E. Gutstein and B. Peterson, pp. 101–2. Milwaukee, Wis.: Rethinking Schools, 2005.

Gutstein, E., and B. Peterson, eds. *Rethinking Mathematics: Teaching Social Justice by the Numbers.* Milwaukee, Wis.: Rethinking Schools, 2005.

Lesh, R., and G. Harel. "Problem Solving, Modeling, and Local Conceptual Development." *Mathematical Thinking and Learning* 5, no. 2–3 (2003): 157–89.

National Council of Teachers of Mathematics (NCTM). *Principles and Standards for School Mathematics.* Reston, Va: NCTM, 2000.

National Research Council (NRC). *Adding It Up: Helping Children Learn Mathematics.* Edited by J. Kilpatrick, J. Swafford, and B. Findell. Washington, D.C.: National Academies Press, 2001.

Stocker, D. *Math That Matters: A Teacher Resource Linking Math and Social Justice.* 2nd ed. Ottawa, ON: Canadian Center for Policy Alternatives, 2008.

Tate, W. F. "Race, Retrenchment, and the Reform of School Mathematics." *Phi Delta Kappan* 75, no. 6 (1994): 477–80.

Use Appropriate Tools Strategically

Practice 5: Use appropriate tools strategically

Mathematically proficient students consider the available tools when solving a mathematical problem. These tools might include pencil and paper, concrete models, a ruler, a protractor, a calculator, a spreadsheet, a computer algebra system, a statistical package, or dynamic geometry software. Proficient students are sufficiently familiar with tools appropriate for their grade or course to make sound decisions about when each of these tools might be helpful, recognizing both the insight to be gained and their limitations. For example, mathematically proficient high school students analyze graphs of functions and solutions generated using a graphing calculator. They detect possible errors by strategically using estimation and other mathematical knowledge. When making mathematical models, they know that technology can enable them to visualize the results of varying assumptions, explore consequences, and compare predictions with data. Mathematically proficient students at various grade levels are able to identify relevant external mathematical resources, such as digital content located on a website, and use them to pose or solve problems. They are able to use technological tools to explore and deepen their understanding of concepts. (CCSSI 2010, p. 7)

Unpacking the Practice

This mathematical practice emphasizes students selecting and using mathematical tools appropriately when engaging in mathematical activities. Students proficient in this practice will be able to consider a tool's usefulness, and its affordances and constraints, as well as know how to use it appropriately. While the main focus of this practice is not foregrounded as a process in the National Council of Teachers of Mathematics' Process Standards (NCTM 2000), there are some clear connections to two of the Standards: Problem Solving and Representation.

Problem Solving Standard

Students should be able to monitor and reflect on their problem-solving process (NCTM 2000, p. 52). As part of this process, students must consider and use tools as a way to understand, represent, and solve the mathematics problems at hand. It is important that teachers introduce students to a wide variety of tools throughout the curriculum. However, it is also important that students learn to use the tools to construct and make sense of concepts for themselves rather than just see teachers using the tools to demonstrate mathematical ideas.

Representation Standard

The kinds of tools in any classroom vary, but the term mathematical tools refers to language, materials, and symbols that can assist students in recording and communicating their methods as well as shaping the way they think about and understand particular mathematical concepts (Hiebert et al. 1997). The Representation Standard states that it is important for pre-K–12 students to be able to do the following: create and use representations to organize, record, and communicate mathematical ideas; select, apply, and translate among math representations to solve problems; and use representations to model and interpret physical, social, and mathematical phenomena (NCTM 2000, p. 67).

Although practice 5 is phrased in terms of what students can do, teachers must play an important role in the development of this practice. NCTM, in the Teaching Principle of *Principles and Standards for School Mathematics* asserts, "students learn mathematics through the experiences that teachers provide. Thus, students' understanding of mathematics, their ability to use it to solve problems, and their confidence in, and disposition toward, mathematics are all shaped by the teaching they encounter in school" (2000, p. 17). Teachers must expose students to various tools, from manipulatives to software, so that students can learn to use these tools to both support and deepen their learning.

In primary grades, students should be encouraged to develop and use tools that make sense to them, often in a way that directly models the task at hand. Teachers should have a variety of materials available so that students can see how different materials might be used to understand or solve mathematics problems. These tools typically include things like pencil and paper to create pictures, concrete manipulatives such as interlocking cubes (e.g., Unifix cubes), base-10 blocks, calculators when appropriate, or even fingers. "A major responsibility of teachers is to create a learning environment in which students' use of multiple representations is encouraged, supported, and accepted by their peers and adults" so that students see the use of tools as appropriate and helpful (NCTM 2000, p. 139).

In upper elementary grades, students should begin to move from concrete materials to more abstract models and representations. This does not mean that concrete materials are not useful or appropriate, but that upper elementary students should be expanding their repertoire by using more sophisticated (and typically more efficient) tools and strategies. In addition, students at this level should be introduced to and have opportunities to use technological tools such as spreadsheets and dynamic geometry software. Teachers should challenge students to develop more than one way to solve problems because "students who represent the problem in more than one way are more likely to see important relationships than those who consider the problem without a representation" (NCTM 2000, p. 206).

As students enter middle school and high school mathematics classes, it becomes more important for them to connect their own ways of representing and solving problems to conventional ways of doing mathematics. This is not to say that secondary students should not have opportunities to invent and use their own methods of solving problems, but that teachers should pose tasks that allow students to make connections between their own strategies and typical ways of using language, symbols, and representations.

At every grade level, teachers play an important role in helping students use tools in ways that clarify the mathematics at hand rather than use tools simply to use them. Thompson (1994) argues that teachers must make judicious and reflective pedagogical decisions about concrete materials, focusing on what they want students to understand rather than what students will learn to do. In other words, a tool like a number line can help elucidate certain ideas about patterns in our number system, but students themselves (with support from teachers and others) construct meanings about numbers. The understanding that students take up may be different than intended. Teachers must also support students in understanding the affordances and constraints of tools, both in relation to specific problems and in general.

Classroom Examples

In these vignettes, we present different ways that students may use mathematical tools as learning supports. In the first example, primary grade children explore multiple solution strategies to solve a problem. In the second example, students use geometry software to assist them in understanding symmetry.

Elementary Grades Vignette: How Many First Graders in Our School?

As Ms. Carr was leading her class through their daily routine of recording lunch and attendance data, first grader Lauren posed the following question to her classmates and teacher: "I know we have 16 kids in our class, but how many first graders are there at Bailey's Elementary?" Although this problem is typically beyond the scope of first-grade curriculum, Ms. Carr believed that her students could solve it with appropriate support. In the following vignette you will see how Ms. Carr supports her students in solving this problem by having them discuss and use a variety of solution strategies and tools.

Ms. Carr: Lauren, that is a very interesting question. How do you think we might be able to figure this out?

Lauren: Well, we could just go around and count each kid...but that might take a long time!

Ms. Carr: You're right, we could go around and count each kid, but it would take a lot of time. Does anyone have another way we could find out this information?

Rajas: We could go in teams to each of the classrooms and count the kids in that room. Then we could report back to the class what we found.

Ms. Carr: Good idea. That might be a little more efficient. Can anyone think of any other ways?

Dani: Well, maybe we could just go ask the teachers in each of the rooms how many students they have. Or we could even just ask Principal Barry.

Ms. Carr: OK, so what do you think we should do, Lauren?

Lauren: I think we should go and ask each teacher to tell us the number of students they have. The principal might know the answer, but might not know which kids are here or absent today.

Ms. Carr: OK, Lauren, why don't you and Dani go to each of the rooms, get that information, and report back to us.

[*A few minutes later, Lauren and Dani return with the information.*]

Ms. Carr: How should we share the numbers with the whole class so that we can begin to figure out the total number of first graders?

Dani: Why don't you just write the numbers on the board.

Ms. Carr: Good idea. I will make a chart.

[*She records the information on the board.*]

[*Although some students might be able to use an algorithm to solve the problem, Ms. Carr writes the equation in figure 5.1 on the board horizontally to encourage her students to use a different strategy.*]

$$16 + 15 + 18 + 15 + 17$$

Fig. 5.1. Sum of first graders in the school

Ms. Carr: OK, so now what do we do?

Rajas: We add all the numbers together.

Ms. Carr: I think that you will be able to solve this problem, but I think we should talk about possible ways to figure this out first.

[Ms. Carr records their ideas (such as base-ten blocks and using a hundreds chart), makes sure that the students have access to necessary materials, and has them work for a few minutes in small groups.]

Ms. Carr: OK, let's come back together to share our results. With any luck, we arrived at the same answer, but I would first like to hear about *how* you solved this problem.

Lauren: Well, the people at my table and I used the base-10 blocks to show each of the classrooms and then we counted up the blocks [*see fig. 5.2*].

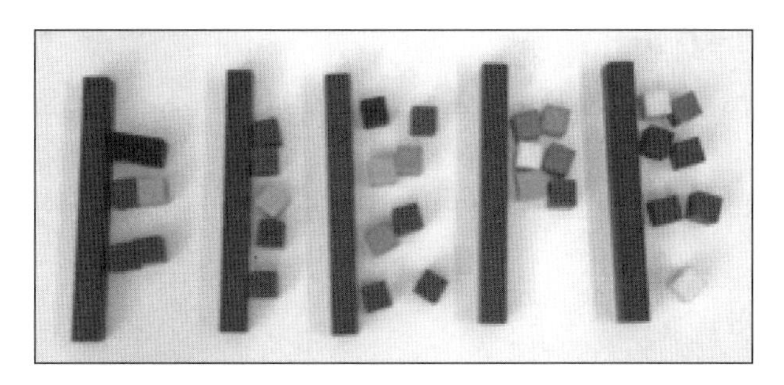

Fig. 5.2. Lauren's initial representation

Ms. Carr: How did you count up the blocks?

Lauren: Well, first, we counted all the 10 rods…10, 20, 30, 40, 50…and then we had to add up all the ones. We knew right away that a group of 5 ones and a group of 5 ones would be another 10, so then we were at 60. Then we just counted the rest by ones.…I think there were 21. So we just counted the 21 on from 60. Our answer is that there are 81 students in all the first-grade classrooms [*see fig. 5.3*].

Fig. 5.3. Lauren's strategy

Ms. Carr:	Lauren, I understand exactly what you did. Great job at explaining your strategy. Does anyone have any questions?
Rajas:	No. We also counted by tens, but we figured it out a lot differently.
Ms. Carr:	OK, tell us how.
Rajas:	We used our hundred charts. First, we started at 16, which is how many kids there are in our class. Then we added 15 more, because that's how many Ms. O'Connor has in her classroom.
Ms. Carr:	Can you show us how you did this? How did you add on 15 on the hundreds grid?
Rajas:	From 16, we jumped down to 26, because that's 10 more. Then we added on the 5. We kept going in the same way for the rest of the classes...adding on 10 and then adding the ones [*see fig. 5.4*].

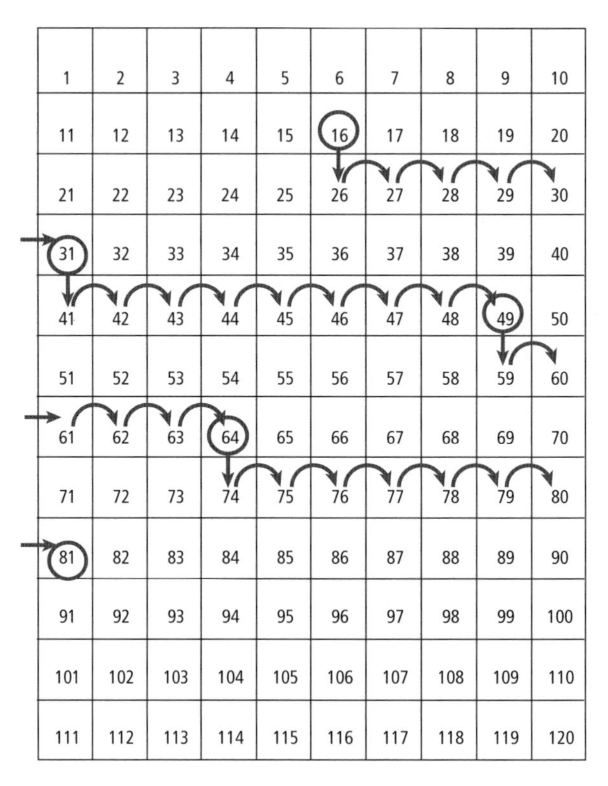

1	2	3	4	5	6	7	8	9	10
11	12	13	14	15	16	17	18	19	20
21	22	23	24	25	26	27	28	29	30
31	32	33	34	35	36	37	38	39	40
41	42	43	44	45	46	47	48	49	50
51	52	53	54	55	56	57	58	59	60
61	62	63	64	65	66	67	68	69	70
71	72	73	74	75	76	77	78	79	80
81	82	83	84	85	86	87	88	89	90
91	92	93	94	95	96	97	98	99	100
101	102	103	104	105	106	107	108	109	110
111	112	113	114	115	116	117	118	119	120

Fig. 5.4. Rajas's strategy

Ms. Carr:	Hmmm. Does anyone have any questions about this group's strategy?
Dani:	No, but we did it two ways. First we used a calculator, but we got it wrong. Then we used a numberline.

Ms. Carr:	How did you know your answer was wrong?
Dani:	Well, we got a number like 100 or something. We knew that we could count by twenties five times and that would be 100. Since all the classes have less than 20 kids, we knew our answer was too much.
Ms. Carr:	Wow, that was really good thinking. Because you estimated, you knew that couldn't be right.
Dani:	You always tell us that we shouldn't just trust calculators...that we need to use our brains! We did use the calculator again, and we thought we got the right answer, but we wanted to make sure and check it. So we also used a number line.

Ms. Carr takes a few minutes to discuss with the class how the hundreds chart strategy and the number line strategy have some similarities. In this discussion, the class also comes to agreement that each of the strategies works, and the answer is 81 students.

In this vignette, it is clear that while the problem would initially be difficult for most first graders, the use of tools facilitated and supported the students as they solved the problem. Presumably the students had access to (and had used) these tools prior to this experience and felt comfortable choosing strategies that they thought would be helpful solving the problem. In the next vignette, you will see how the use of software can support students in more fully understanding a problem.

Middle School Vignette: Rotations and Reflections

In the following vignette, which draws heavily from Andrea B. Graf's (2010) article "Think Outside the Polygon," students engage in several spatial visualization tasks that are enhanced by the use of dynamic geometry software, such as Geometer's Sketchpad or GeoGebra. This vignette is of particular value because it highlights how mathematical tools can help provide insight into the mathematical concepts at hand—in this case the effects of rotating or reflecting figures in the Cartesian plane.

Ms. Graf's students have studied lines of symmetry and points of rotational symmetry within polygons, but now Ms. Graf wants to push them to consider what happens when polygons are reflected and rotated in other ways. Based on her past experience with this lesson, she knows that a number of students struggle to engage with the mathematical substance of reflections and rotations because of their lack of drawing experience, so Ms. Graf uses Geometer's Sketchpad to help with drawing the figures. She uses the software to display the coordinate grid, has the program "snap" the points to the coordinate grid to make the creation of polygons easier, and creates an asymmetric polygon in the first quadrant (see fig. 5.5). The class works together to reflect the image into the remaining three quadrants.

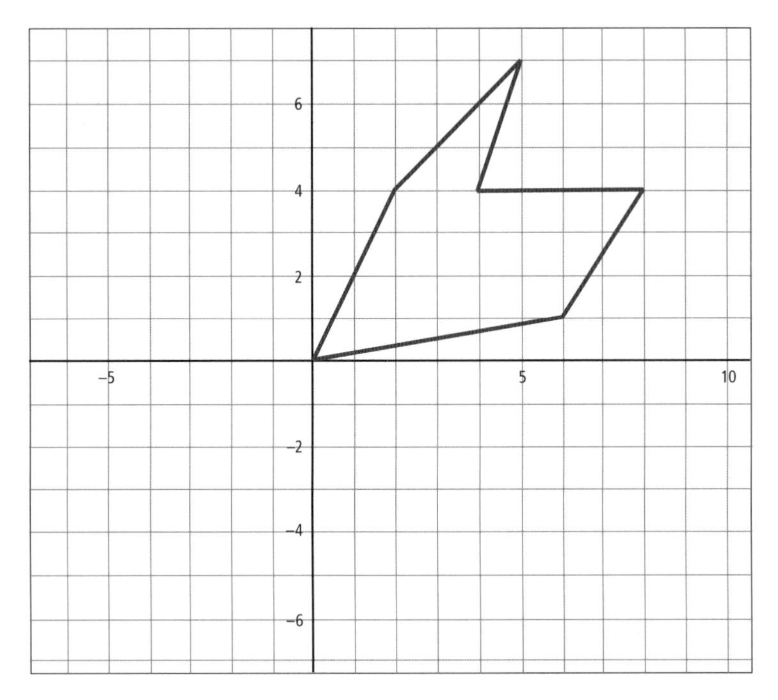

Fig. 5.5. Polygon in first quadrant (re-created from Graf [2010, p. 84])

Ms. Graf: OK, so where will this polygon end up if I reflect it over the y-axis?

Finn: It will end up over here, on the left.

Ms. Graf: Can you come show the class where you think it will end up?

[*Finn comes to the board and points to the second quadrant.*]

Finn: It will be here, still pointing into the middle, but flipped over.

Ms. Graf: OK, great, so what we're going to do is break this down piece by piece. We'll just reflect one line of the polygon at a time and figure out where it will end up. Can someone else come up and show me where this line segment [*points to the segment from (0, 0) to (2, 4)*] will end up after the reflection?

[*A student comes up and points to one end point of (0, 0) and another end point of (−2, 4). This process is repeated for several edges of the polygon. The students label the end points of each edge as they go.*]

Ms. Graf: Alright, we've done a bunch of these now. I want you to turn to your neighbor and discuss how the new points, after the reflection, are labeled compared to where they started out.

[Several students in the class observe that the coordinates change from (x, y) to (−x, y). Ms. Graf then continues the lesson as a whole class, reflecting the rest of the edges into the second quadrant and then repeating this process by reflecting over the fourth quadrant. Throughout this process she graphs both correct and incorrect answers from students and allows the class to make corrections based on their visual inspection or consideration of the coordinates. In working on the reflection over the x-axis, the students soon discover that the coordinates shift from (x, y) to (x, −y). Finally, Ms. Graf has the class reflect the shape into the third quadrant.]

Ms. Graf: OK, now let's reflect this into the third quadrant. What should we do?

Josue: Well, do we start from the top one [*quadrant 2*] or from the one on the right [*quadrant 4*]?

Ms. Graf: Let's try them both out and see?

[Students try translating a corresponding segment first from the second quadrant and then from the fourth quadrant and quickly conclude that it does not matter.]

Ms. Graf: Thinking about what we've done so far, why would it turn out the same either way? Is there any way we can tell if this will always work?

Hayley: I think I know. [*Comes up to board to share.*] If you reflect it one way, like this [*indicating a reflection over the y-axis*] then you change the coordinates to negative x, y. Then if you reflect it down like this [*indicating a reflection of the x-axis*] you will make the y negative. So it will be negative x, negative y. If you go the other way the same thing happens just in the other order.

The class finishes creating the reflected polygon in the third quadrant. Then Ms. Graf has each student work on their own computers to create the polygon in the first quadrant and reflect it into the other three quadrants. The next day this activity is extended to consider what would happen if the same figure were rotated around the origin (instead of reflected over the axes) into each new quadrant. Ms. Graf uses a cardboard cutout of the polygon that she can physically rotate around the origin to help with the visualization.

Although tasks such as those described above could be solved without the use of computer software, Graf (2010) asserts that dynamic geometry software can eliminate the frustrations of students who have a difficult time with visual and spatial tasks because of their lack of experience with drawing. The vignette above shows how a teacher can selectively use mathematical tools to aid students in learning important mathematics. The vignette also shows how Ms. Graf is able to use the mathematical tool to help students avoid the tedious and difficult aspects of the work (drawing, erasing, and redrawing messy figures) and instead focus on the mathematics involved. However, she is also careful in how she uses the tools. The software could perform the reflections and rotations automatically, but by not using this part of the software, the students are still doing the relevant mathematical work. Finally, when needed, she supplements one tool (the software) with another (the cardboard cutout) to help students visualize the geometric transformations in another way.

Resources

This resource explores the use of tools among primary students during problem solving and shows how using tools can support students in understanding and representing mathematics problems.

- Jacobs, V. R., and J. Kusiak. "Got Tools? Exploring Children's Use of Mathematics Tools during Problem Solving." *Teaching Children Mathematics* 12, no. 9 (2006): 470–77.

This resource provides an overview of virtual manipulatives and how they might be used in mathematics classrooms.

- Moyer, P., J. Bolyard, and M. Spikell. "What Are Virtual Manipulatives?" *Teaching Children Mathematics* 8, no. 6 (2002): 372.

This resource shows how teachers created technology-rich environments. The article provides examples of how tools were used to support diverse students in learning mathematics.

- Suh, J. M., C. J. Johnston, and J. Douds. "Enhancing Mathematics Learning in a Technology-Rich Environment." *Teaching Children Mathematics* 14, no. 4 (2008): 235–41.

The resource argues and gives justification for the thoughtful use of manipulatives in the middle grades.

- Weiss, D. F. "Keeping It Real: The Rationale for Using Manipulatives in the Middle Grades." *Mathematics Teaching in the Middle School* 11, no. 5 (2006): 238–42.

This resource shows how nonroutine problems can encourage and support students in developing, using, and comparing multiple solution strategies and representations, including the use of graphing calculators.

- Trinter, C. P., and J. Garofalo. "Exploring Nonroutine Functions Algebraically and Graphically. *The Mathematics Teacher* 104, no. 7 (2011): 508–13.

This resource offers a vignette that describes how students might solve open-ended problems using multiple technologies.

- Erbas, A. K., S. D. Ledford, C. Orrill, and D. Polly. "Promoting Problem Solving across Geometry and Algebra." *The Mathematics Teacher* 98, no. 9 (2005): 599–603.

References

Common Core State Standards Initiative (CCSSI). *Common Core State Standards for Mathematics.* Washington, D.C.: National Governors Association Center for Best Practices and the Council of Chief State School Officers, 2010. http://www.corestandards.org.

Graf, A. B. "Think Outside the Polygon." *Mathematics Teaching in the Middle School* 16, no. 2 (2010): 82–87.

Hiebert, J., T. P. Carpenter, E. Fennema, K. C. Fuson, D. Wearne, H. Murray, A. Olivier, and P. Human. *Making Sense: Teaching and Learning Mathematics with Understanding.* Portsmouth, N.H.: Heinemann, 1997.

National Council of Teachers of Mathematics (NCTM). *Principles and Standards for School Mathematics.* Reston, Va: NCTM, 2000.

Thompson, P. "Concrete Materials and Teaching for Mathematical Understanding." *The Arithmetic Teacher* 41, no. 9 (1994): 556–58.

Attend to Precision

Practice 6: Attend to precision

Mathematically proficient students try to communicate precisely to others. They try to use clear definitions in discussion with others and in their own reasoning. They state the meaning of the symbols they choose, including using the equal sign consistently and appropriately. They are careful about specifying units of measure, and labeling axes to clarify the correspondence with quantities in a problem. They calculate accurately and efficiently, express numerical answers with a degree of precision appropriate for the problem context. In the elementary grades, students give carefully formulated explanations to each other. By the time they reach high school they have learned to examine claims and make explicit use of definitions. (CCSSI 2010, p. 7)

Unpacking the Practice

An important characteristic of mathematics is precision of language. In mathematics, we try to consistently say what we mean and mean what we say. As students move through school mathematics, it is important for them to come to understand and be able to communicate with precise language as they use definitions, formulate explanations, make and refine conjectures, and construct and critique mathematical arguments. All these activities depend on the precise use of language. There are others ways in which precision is important, as well. For example, in measurement, one should always be aware of the level of precision required in a given situation and the level of precision appropriate for a given instrument. When performing numerical calculations, decisions about rounding should be made by thoughtfully considering precision. Precision can also be an important factor in mathematical thinking. As an example, consider a student who is attempting to determine whether $^{23}/_{24}$ is the sum of $^{7}/_{12}$ and $^{5}/_{8}$. At one level of precision, this answer seems reasonable because $^{7}/_{12}$ and $^{5}/_{8}$ are both approximately equal to $^{1}/_{2}$ and $^{23}/_{24}$ is approximately equal to 1. At another level of precision, however, one can conclude that $23/24$ is not the correct sum because $23/24$ is less than 1, whereas $7/12$ and $5/8$ are both greater than $1/2$.

In this chapter, we first unpack ideas concerning precision of language and then move on to consider numerical precision, connecting these aspects of the mathematical practice to the National Council of Teachers of Mathematics' (NCTM's) Communication and Representation Process Standards, respectively. Then, we present examples of the practice of attending to precision in the elementary and secondary levels. NCTM resources related to this practice can be found at the end of the chapter.

Communication Standard

As students progress through their school mathematics experiences, our goal as teachers is not only for them to develop a set of mathematical skills and acquire a wealth of mathematical knowledge but also for them to become capable communicators of mathematical ideas. Within the classroom community, this communication might take the form of students showing their work so that classmates or the teacher can follow the reasoning behind a solution. It may also take the form of students explaining their thinking in small groups, asking coherent questions during a whole-class discussion, or writing solutions for the teacher to read. As noted in NCTM's (2000) Communication Standard in *Principles and Standards for School Mathematics*, there is an important balance to strike between idiosyncrasies arising in student communication and the conventional standards of communication in the broader mathematical community. In particular, idiosyncrasies in communication arising from students' prior knowledge or their own organic interpretations of mathematical phenomena can be powerful foundations for conceptual understanding. However, it is also important for teachers to help students learn the conventional modes of communication so that students may be viewed as capable communicators in broader contexts such as statewide assessments, university mathematics departments, or the job market.

Standards of mathematical communication involve vocabulary, accepted forms of argumentation, conventions for symbolic and graphical representations, and many other features. Here we focus on the role of precision in mathematical communication and note the connections as well as the distinctions between precision and formality of language.

Mathematical definitions. Within the Communication Standard, NCTM (2000) marked the importance of students being able to "use the language of mathematics to express mathematical ideas precisely" (p. 60) and having "experiences that help them appreciate the power and precision of mathematical language" (p. 63). Much of this "power and precision" begins with mathematical definitions, which are qualitatively different than definitions in areas other than mathematics. In the everyday, dictionary sense, definitions are meticulous descriptions of the meaning of words as they are used in a particular language. For example, one definition of the noun "ball" is a spherical or approximately spherical body or shape. One can imagine that this description arose as an attempt to capture many of the ways in which the word "ball" was being used, from beach balls (spherical) to footballs (approximately spherical) to crumpled pieces of paper or balls of cookie dough. An important point to note is that the definition was formed after and in response to usage of the term. Similarly, many scientific definitions are careful characterizations of observed phenomena in the world.

Definitions in mathematics function quite differently. The defined mathematical objects, in all instances, have precisely the characteristics specified by their definitions and, furthermore, any object with those specified characteristics is necessarily an instance of that mathematical object. It is this fact that grants mathematical terms a much higher degree of precision than students are likely to experience in other settings. For example, a light bulb and a soap bubble fit the definition of "ball" given above but we can agree that these are not balls, which implies that the definition is not perfectly precise. It also implies that there is a certain element of "knowing it when you see it" involved in our determination of what is and is not a ball. In mathematics, however, any quadrilateral with four congruent sides is, *by definition*, a rhombus, and any nonquadrilateral or quadrilateral without all sides congruent is not a rhombus. The definition is precise.

Why does all this matter for students? It matters because students come to the mathematics classroom with ample experience using everyday language, where definitions are typically viewed as subordinate to knowing it when you see it. If they are not aware of the shift in precision in the mathematics classroom, elementary students may think of triangles as only including equilateral or right triangles, and secondary students may conceptualize function as any relation between variables rather than a relation in which each input is associated with exactly one output. Over time, teachers can build on these intuitive judgments toward more precise mathematical definitions by asking students to articulate their concept images, by having students explore examples and nonexamples of mathematical objects, and by explicitly addressing characteristics of a good definition and the reasons for those characteristics.

We reiterate that attending to precision is a process of building on students' informal knowledge, not a replacement of it. As NCTM (2000) put it:

> Students need to develop an appreciation of the need for precise definitions and for the communicative power of conventional mathematical terms by first communicating in their own words. Allowing students to grapple with their ideas and develop their own informal means of expressing them can be an effective way to foster engagement and ownership. (p. 63)

This passage also indicates that a focus on definitions does not mean an increase in vocabulary assignments or requirements that students memorize definitions word for word. Moschkovich (2007), in particular, points out that an overemphasis on technical vocabulary can be detrimental to students' learning, especially in English language learner contexts. Rather, the goal is to help students appreciate the role of definitions in the practice of doing mathematics. In fact, *defining* is an integral part of the reasoning-and-proving process and the creation of mathematical knowledge (Lakatos 1976). When teachers engage students actively in the formulation of mathematical definitions, students may come to both understand the concepts associated with those definitions and appreciate the role of definitions in mathematical communication.

Another way that teachers can create opportunities for students to attend to the precision of mathematical definitions is by having them reflect explicitly on the fact that "some words that are used in everyday language, such as *similar, factor, area*, or *function*, are used in mathematics with different or more-precise meanings" (NCTM 2000, p. 63). The next section addresses some of the other ways in which mathematical language is different or more precise than everyday language.

Mathematical grammar. In addition to the precise terminology of mathematics, there is also precision in the grammar—the way words are put together and used to express meaning. Again, the point is not to drill perfect grammar into students from day one but to support students over time so that they may become effective communicators in the classroom community as well as the broader mathematical community, recognizing that "reflection and communication are intertwined processes in mathematics learning" (NCTM 2000, p. 61).

Within mathematics communities, quantifiers (e.g., every, for all, there exists), negations, and logical connectors (e.g., so, then, because) all take on precise meanings, beyond the ways they might be used in everyday settings (Schleppegrell 2007). For example, in mathematics when we talk of something being true *in general*, we do not mean that it holds on average or roughly speaking. Additionally, if a teacher proposes a conjecture about *all* even numbers, it must hold for literally every even number. The precision of this statement has important implications for the reasoning that will be entailed in proving or disproving the conjecture. In some cases, students may feel overburdened by the task of showing something to be true for all members of an infinite set, but this seemingly daunting task becomes feasible when one realizes that, as discussed above, mathematical definitions completely determine and demarcate the set in question.

As students explain their thinking and reasoning in mathematics classes, they are likely to begin by recounting the actions they took in solving the problem. For example, a student who solved $22 = 5x + 2$ might explain, "I started by subtracting two from both sides, and then I divided by the five to get x by itself, so I got x equals four." In this explanation, *then* and *so* are used to mark chronology. Although there is nothing wrong with these words being used in this way, it is important for students to gradually come to also use these words to articulate mathematical relationships and logical dependencies. In particular, *then* can be a significant marker of a conditional relationship and *so* can be used in a chain of reasoning as a synonym for *thus* or *therefore*. Another key mathematical term is *because*, which teachers can encourage students to use by asking them *why* questions. Overall, by gaining experience using these words to express their mathematical thinking, students can learn to precisely explain not only what they did but the connections and relationships that underlie their work.

Another important aspect of mathematical grammar has to do with symbolic and graphical representations. Mathematical communication relies on more than just words to carry meaning—symbols, graphs, tables, and even gestures are valuable resources for communication in mathematics. These meaning-making resources can be used in powerful ways to express mathematical ideas and have the potential, as some of the classroom examples below demonstrate, to represent an idea even more clearly than words can. But just like with words, teachers should support students in attending to precision when they use these other resources to communicate. For example, variables should be defined when they are introduced into a problem, graphs should be labeled, quantities should be expressed with units, and symbols such as "=" should be used in ways that agree with the broader community's usage (e.g., not as in $5 + 7 = 12 \div 6 = 2$). Students should also understand the difference in precision between a geometric sketch and a geometric construction (although the latter is still not perfectly precise as the use of tools introduces some error) as well as between a simple sketch and a sketch with markings such as right angles and congruent sides. By attending to the precision of their mathematical communications, whether spoken, written, or sketched, students can maximize the opportunities for others to understand their ideas. In this way, attention to precision in communication can be the basis of many good mathematical conversations.

Precision and formality. In an effort to be precise in our own discussion of precision in language, we wish to clarify that, with respect to language, we have been using the word *precision* to refer to the goal of meaning what we say (or write) and saying (or writing) what we mean. This notion of precision is related to but distinct from the *formality* of mathematical language. Precision and formality are linked in the sense that a desire to communicate precisely in mathematics often leads one to use formal language, recognizing that this formal language in many cases has been developed specifically to allow for precise communication. However, it is possible for informal language to be perfectly precise. For example, if a student has the line $y = x$ graphed on a piece of graph paper and gestures along it from left to right as she says, "It's going up," then there is little ambiguity in what this statement means. Clearly, *it* refers to the graph of the line and *going up* can be interpreted as meaning that the y-values are rising from left to right because this was the direction of her gesture. Although this communication is precise, its meaning depends on several other features of the situation, not solely on the words spoken. A more formal phrasing, such as "if x and y are related by the equation $y = x$, then larger values of x are associated with larger values of y," makes it possible to express the same meaning without relying on situational features such as the graph or gestures. Having the ability to communicate in this formal manner would be beneficial for students, but it is not necessary or efficient to talk in this way at all times—if situational features are available, we should expect students to use them when communicating (Gibbons 2009). In other words, precision can be attended to in many different contexts of communication but formal language may be more or less appropriate in different contexts, and students' facility with formal language can be a long-term learning goal. NCTM (2000) made a similar point when articulating its Communication Standard:

> For some purposes it will be appropriate for students to describe their thinking informally, using ordinary language and sketches, but they should also learn to communicate in more-formal mathematical ways, using conventional mathematical terminology, through the middle grades and into high school. By the end of the high school years, students should be able to write well-constructed mathematical arguments using formal vocabulary. (p. 62)

Representation Standard

Just as mathematical definitions are precise in ways that definitions in other areas are not, mathematics calculations are often exact rather than approximate, and so it becomes important for students to recognize how and when exactness plays a role in their mathematical work. Furthermore, mathematics, especially in the early grades, is the place where an understanding of the decimal number system is explicitly developed, which has significant implications with respect to precision. The Representation Standard from *Principles and Standards for School Mathematics* (NCTM 2000) includes students' facility with decimal numbers as well as other expressions of number that have implications for precision, such as fractions and symbols (e.g., π, $\sqrt{}$).

Precision with decimal numbers. The structure of the decimal number system allows for great flexibility when it comes to precision. Within the natural numbers, students can round to the nearest ten, hundred, thousand, and so forth, as appropriate. They can also gain experience in estimating the result of numerical computations and discuss the benefits and limitations of these estimates. As NCTM (2000) has pointed out, "Estimation serves as an important companion to computation. It provides a tool for judging the reasonableness of calculator, mental, and paper-and-pencil computations" (p. 155). It is important, however, that such estimation practices occur together with an understanding of the implications for precision.

Additionally, with respect to the decimal number system, students can view tenths, hundredths, thousandths, and so forth, as progressively more precise ways to represent a number or quantity. Especially in the case of repeating decimals or nonrepeating decimals, students should develop an awareness of the loss of precision that occurs through truncation or rounding and should come to realize that the imprecision is carried forward with the rounded number. An especially powerful way to explore these ideas is through measurement, which can allow for connections to other disciplines, such as science, where precision is also a key concept. Although one cannot be perfectly precise in realistic measurement contexts, it is important to attend to the level of precision that is achieved, which connects to the notion of "significant digits."

Especially in the elementary grades, where developing a strong understanding of decimal numbers and the place-value system is central in both *Principles and Standards* (NCTM 2000) and the *Common Core State Standards for Mathematics* (CCSSM; CCSSI 2010), attending to precision and learning the decimal numbers as a system can be mutually supportive.

Precision in calculations. When solving a purely mathematical task, there is no reason to water down exact solutions to approximations except that the approximations, which often come in the form of rounded decimal numbers, may be more familiar or comfortable to the students. Part of the mathematical practice of attending to precision, then, is to see the value in and become comfortable with exact solutions. Once students learn the decimal number system, you may notice them converting fractions to decimals automatically, almost on instinct. Some reasons for this might be that decimal numbers give a clearer sense of the size of a number than fractions (e.g., 13.125 vs. $^{735}/_{56}$), decimal numbers are more common to students' out-of-school experiences, and decimal numbers are more amenable to calculator use. Regardless of the reasons for favoring decimals, it is important for students to realize that any rounding when they convert to decimals results in a loss of precision.

Later on, when irrational numbers come into the picture, conversions to decimal numbers unavoidably involve a loss of precision. In fact, symbols such as π and e originated from an attention to precision because these symbols are perfectly precise, and without them, decimal approximations or infinite series would be needed to represent these numbers. Thus, it is beneficial for students to recognize that decimals are not the only legitimate type of numerical representation. Students who display a strong urge to give the decimal approximation as their final answer, especially when the exact solution involves a fraction bar or a radical, may be revealing that they view these symbols as indicating an operation to perform (i.e., "dividing" and "taking the square root," respectively) rather than viewing them as mathematical objects in and of themselves. Helping these students attend to precision may provide an avenue for moving them to the next level of mathematical understanding.

Connections to Other Mathematical Practices

The ideas related to numerical precision described above are not about following the directions of a teacher or textbook about when and how to round. They are an important aspect of number sense—recognizing when approximation may be appropriate and knowing approximation when you see it, but also knowing when there is nothing gained by approximation and it would be wiser to keep things exact. These ideas tie in with quantitative literacy (see practice 2), which highlights the fact that it is valuable to attend to precision outside of mathematics class as well. For example, students with a keen understanding of precision may notice in their everyday life that approximations and calculations based on imperfect measurements are often presented as factual figures.

Modeling with mathematics (practice 4) highlights the role that mathematics plays in representing and solving problems from everyday life as well as other disciplines such as science and engineering. When mathematics is used in such domains, which are not purely mathematical, some of the messiness and imprecision of the world creeps into the work, as it should. Managing this messiness in productive ways is an important part of the process of modeling, which implies that the practice of attending to precision should always be close at hand when one is modeling.

Another practice with strong connections to attending to precision, mentioned earlier in this chapter, is the construction and critique of arguments (practice 3). In particular, definitions are an important component of reasoning and proof, as the development of a definition often occurs hand-in-hand with reasoning about a concept, and definitions are extremely useful in proof arguments. Moreover, the practice of critiquing the reasoning of others often involves at least two dimensions: the first relating to critique of underlying reasons and logic and the second relating to critique of the expression of the ideas in an argument. For example, a student may challenge another student's proof by pointing out that it actually proved something different than the original claim or that it cited an inappropriate justification. Alternatively, a student may challenge another student's proof by identifying ideas within the proof that are not clear or are not phrased properly, even though the underlying idea is valid. The latter critique often involves attention to precision that, as a mathematical practice, can occur in a mutually beneficial way with reasoning and proof.

Finally, whenever tools are used (practice 5), whether they are digital (e.g., calculators, applets) or physical (e.g., rulers, protractors), students should exercise an awareness of the level of precision involved. This attention to precision is an important component of appropriate tool use.

Classroom Examples

Because the practice of attending to precision encompasses several facets of mathematical activity, from precision in language to precision in calculations and measurement, this section includes several practical examples at both the elementary and secondary levels. Each example is meant to illuminate a different aspect of the practice, and taken together they highlight the fact that attention to precision should always be welcome in mathematics classrooms.

Elementary Grades Vignettes

Defining geometric shapes. Understanding the various relationships between two-dimensional shapes is an important part of the geometry strand in the elementary grades. This understanding of the relationships between shapes can develop with and be mutually supportive of students' understanding of individual shapes in and of themselves. Over time, students should move from recognizing certain shapes only by visual cues, which sometimes leaves out unconventional versions of shapes (e.g., obtuse triangles, extremely skinny rectangles, squares sitting on their corner), to being able to articulate definitions of shapes and determine examples and nonexamples from those definitions. Using the precise definitions of shapes is also a mathematically appropriate way to settle disputes such as whether or not a square is a rectangle.

Ms. Jackson: Is a square a rectangle? Some of you are saying yes and some are saying no. Let's try to figure this out. Michael, what was it that you were saying again?

Michael: I think it's a rectangle.

Ms. Jackson: And why do you think a square is a rectangle?

Michael: Because it looks like one. I mean, if I didn't know for sure that it was a square, I would definitely say it's a rectangle.

Ms. Jackson: And Darlene, would you mind repeating what you were saying earlier so we can think more about it?

Darlene: I don't think it's a rectangle because it's a square, so it already has a shape.

Ms. Jackson: OK, so we have a couple different ideas to think about. I want to add another idea into the mix. I want us to think about exactly what a rectangle is, OK? Remember what we learned about rectangles. There's something special about their angles. What is it?

Students: Right angles.

Ms. Jackson: Yeah, all four angles are right angles. That's what makes a shape a rectangle. Now, let's look at our square [*pointing to the board*]. What can you see about its angles? Julia?

Julia: They're all right.

Ms. Jackson: They're right, aren't they? [*She gestures to each of the square's angles.*] And any four-sided shape with four right angles is a rectangle, so this is a rectangle. And Michael, you were saying that it looked like a rectangle. That's because it *is* a rectangle. It's a square *and* a rectangle. Darlene, you were saying that it's a square, so it can't be a rectangle, because each shape only has one name. But actually, shapes can be more than one thing at the same time! This shape is a square, but it's also a rectangle because it has four right angles. It's both. Just like you can be more than one thing at the same time—you're a student, a daughter, a girl, a human being, right? You're all those things. And this shape is a square and a rectangle, and we'll even learn later that it's a few other things, too.

The students came to this problem with strategies for answering the question about whether a square can also be a rectangle, but their notions of the shapes were more intuitive and imprecise. Ms. Jackson steered the discussion in a more precise direction, using the definition of rectangle to come to a conclusive answer to the question that could otherwise seem like a matter of opinion. In this example, the teacher modeled attention to precision in a way that helped students make progress on the problem. It is not necessarily clear, however, if all the students were on board with her answer to the question, which highlights the fact that this process of attending to precise definitions takes place over time. In the future, she can work to relinquish more of the control to students as they become more comfortable in attending to precision, seeing its value and power.

Numerical estimation tasks. Rounding and estimation are useful skills for young students to develop, but it is important that students recognize when they are estimating and when they are working with precise values. Tasks such as the one in figure 6.1 not only give students opportunities to develop their numerical skills, but can also open up discussions of the precision of their work.

Pick three problems and estimate the answer. For the other three problems, find the exact answer.		
a) 19 + 22	b) 11 × 4	c) 103 – 51
d) 81 ÷ 20	e) 38 – 18	f) 3 × 1003

Fig. 6.1. Estimation vs. exact answer task

After working on this task, students can discuss their work with their classmates. For instance, students who estimated certain problems can compare their work with other students who found the exact answer. Students can also talk about why they chose the problems they did to estimate. The existence of that choice can bring the issue of precision to the surface.

Later on, students can think about fractional representations of rational numbers versus decimal representations. Many students feel more comfortable with decimal representations, perhaps because it gives a clearer sense of the magnitude of a number (e.g., about how big is $^{2329}/_{107}$? 21.766?). But proficient mathematical thinkers often prefer rational representations because they are perfectly precise (and because they sometimes reveal structure; see practice 7). Older students who are quick to convert rational numbers to decimals can be asked why they make those choices, and this conversation can be used to encourage students to attend to the precision of their work.

Precision and context. Versions of the following task are well-known (e.g., Silver, Shapiro, and Deutsch 1993) because of their appearance on standardized tests and because of the issues they raise with respect to mathematical solutions and problem contexts:

> The sixth grade at Hickory Middle School is taking a field trip. There are 373 people, including students, teachers, and chaperones. They will travel by bus, and each bus holds 40 people. How many buses will they need for the field trip?

This task can also be viewed through the lens of precision—how much precision is appropriate given a particular problem context? Students often determine the numerical answer of 9.325, but this level of precision is not appropriate for the way the problem is phrased, which implies that only whole buses can be used. In this case, attending to precision involves looking closely at the specific parameters of the problem as well as attending to the exactness of the answer, realizing that 10 is a more appropriate response than 9.325 (even though the latter is more computationally precise). Moreover, collectively attending to precision might also open the door to more creative responses, which, if accompanied by coherent rationales, can also be solutions to the problem. For example, students might suggest that 9 buses would suffice with the smallest 13 students squeezed in with other students or that some of the chaperones have to provide their own transportation or that 9 buses can be reserved along with 1 shuttle or van.

Precision and measurement. The idea of repeatedly measuring objects arises in the *Common Core State Standards for Mathematics* in the content standards, but it can also be related to the mathematical practice of attending to precision. In particular, second graders are expected to measure the same object with different-sized units and then describe the relationship between the measurements and the size of the unit (CCSSI 2010, 2.MD.2). A lesson geared toward this content standard can also engage students in attending to precision because different-sized units can be interpreted as measurement tools with different levels of precision. Other CCSSM content standards (2.MD.9; 3.MD.4) also involve repeated measurements for the purpose of populating a data set. Whenever an object is being measured multiple times, the topic of precision can arise. Even if students use the same standard unit each time, perhaps the first measurement is done to the nearest whole inch (or centimeter), the second to the nearest half inch (or half centimeter), and the third to the nearest quarter inch (or tenth of a centimeter). This process might help students realize that all physical measurements are approximations and thus not perfectly precise.

Middle and High School Vignettes

Developing a definition. The first elementary example showed how mathematical definitions can be used to answer important mathematical questions. Students can also be actively involved in the process of generating and refining mathematical definitions. This defining process requires attention to precision and is also intertwined with the broader practice of constructing arguments and justifications (i.e., claims and proofs depend on the definitions of the terms involved). In the excerpt below, which comes from a high school geometry class, the teacher and his students are working to come to agreement on the definition of a kite. The class had previously defined several shapes, and the teacher, Mr. Green, asked students to quickly write down their initial definitions of kite. Having briefly looked at a variety of students' responses, Mr. Green has chosen a few as the basis for starting a discussion that will lead to a precise definition of kite.

Mr. Green: Cassie, would you mind sharing your initial thoughts?

Cassie: Um, I had that a kite is a shape with two sets of equal sides.

Mr. Green: OK [*records Cassie's initial definition on the board (see fig. 6.2)*]. Did others have this definition, or something similar to this, about sets of sides being equal? [*Several students raise their hands.*] OK, so let's think about this definition. Here are a few kites, the things that we're trying to define [*draws two kites on the board (see fig. 6.2)*]. Does this definition include these kites?

Students: Yeah.

A kite is a shape with two sets of equal sides

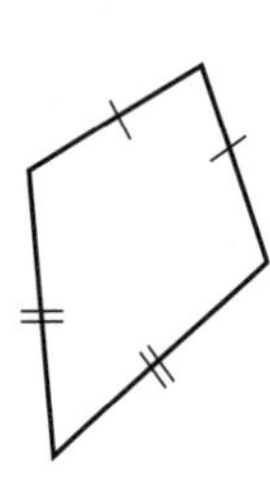

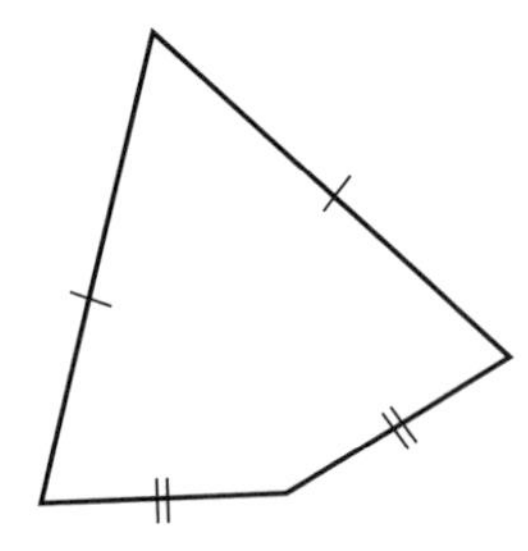

Fig. 6.2. Trying to define kites

Mr. Green: Yes it does. In fact, Cassie's definition is going to include all kites because all kites have two sets of equal sides, and that's a good thing because we want the definition to capture the thing that we're trying to define. But…

Nadia: But what about a rectangle.

Mr. Green: What about a rectangle?

Nadia: Well, a rectangle has two sets of equal sides but it's not supposed to be a kite.

Brendan: And a parallelogram.

Mr. Green: That's an interesting point. A rectangle [*draws a rectangle on the board*] has two pairs of equal sides so it fits our working definition, but we do not want to include it as a kite. So our working definition is a good start, but we need to refine it a bit. Juan, would you mind sharing what you wrote down?

Juan: [*Reading from his paper*] A kite has two sets of consecutive equal sides.

Mr. Green: [*Recording on the board*]…consecutive equal sides. You might also say "adjacent equal sides," that would mean the same thing. OK. This still works as a description of the kites that we want [*gestures to the kites*], but it rules out rectangles because a rectangle's equal sides are not adjacent, they are opposite [*gestures to the rectangle*]. So let's take this to be our new working definition. Remember, we want our definition to capture all the kites and nothing else, no other shapes. So now I want you to take a minute or so to think about this new working definition, and you can keep thinking about what you wrote down earlier, too. Is this definition good to go, or are there still shapes slipping in there that we don't want to include? Take about a minute to think, and feel free to draw on your paper as you're thinking. [*Students work for several moments.*] OK, I saw some good thinking going on, and I've asked Tanya to draw her shape on the board so we can talk about it. Tanya, tell us what you were thinking about this shape.

Tanya: Well, this shape [*fig. 6.3*] has two sets of equal sides but it's not a kite because it has five total sides.

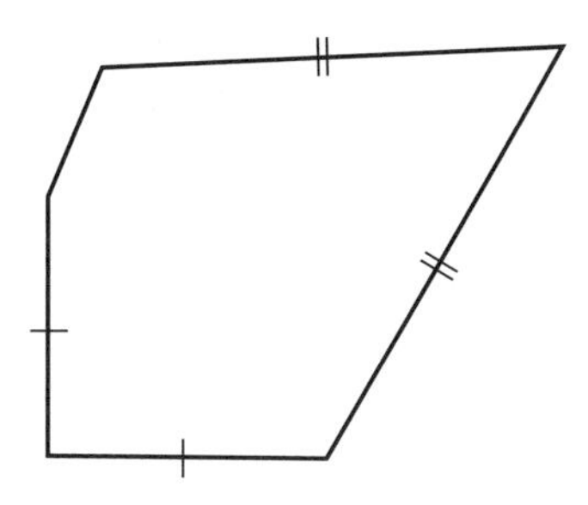

Fig. 6.3. A non-kite

Mr. Green: So you're saying our current version of the definition is not precise enough yet, because you still found a shape that fits the definition—two sets of consecutive equal sides [*gestures to the shape*]—but is not actually a kite.

Tanya: Yeah.

Mr. Green: Does anyone have some thinking to share that helps us get past this problem Tanya has raised?

Curtis: I think it has to have four sides.

Mr. Green: And how did you write that into your definition?

Curtis: I just said it has to have four sides, and then I said the part about the consecutive equal sides.

Mr. Green: What is our official term for a polygon with four sides?

Students: Quadrilateral.

Mr. Green: Quadrilateral, right? We defined that last time, so we can go ahead and use it in our definition. [*Records on the board as he speaks.*] A kite is a quadrilateral—it has to have four sides—with two sets of consecutive equal sides. So Tanya's figure has helped us improve our definition because now we're sure to say that a kite has four sides. In fact, that happens in many definitions in mathematics—you start by saying what your thing actually is, in our case, it's a quadrilateral, and then you go on to say what the special characteristics are of your thing. For us, it's the consecutive equal sides that make a kite a kite.

In this example, the teacher did not simply present the definition to students in finished form but instead used the development of the definition as an opportunity to work with students in attending to the precision of the language in the definition (and the potential definitions). This work also allowed students to connect and distinguish kites from other shapes and gave students a chance to try to think of exceptions to a statement, another good mathematical skill to develop.

It should be noted that the process being undertaken in Mr. Green's class was a refinement of the definition of kite from something that was too broad to something that accurately captured the concept of kite. Careful attention to precision can guide such a refinement, but it is also possible for attention to precision to guide a process of loosening restrictions. For example, the students might have started with a definition of kite that was too narrow and then worked together to broaden their definition appropriately. Moving in either direction involves attending to precision.

Language in reasoning. Attending to precision is inseparable from mathematical reasoning and explanation because we use language to communicate our mathematical thinking, and imprecise language may obfuscate what we are actually thinking. In the next classroom excerpt, a student shares an idea that is not wrong, and in fact it is quite easy to tell what the student means, but the teacher uses this as an opportunity to push on the language being used, thus improving the precision of the ideas being shared. Eventually, this process of pushing on the use of language leads to a more detailed look at what is going on with the two functions.

The problem being considered in this excerpt involves making a choice about where to invest \$2000 in savings. Account A yields linear growth over t years (i.e., $f(\mathrm{t}) = 2000 + 120t$), whereas Account B yields exponential growth (i.e., $g(t) = 2000\ (1.05)^t$). Students are supposed to base their decision of which account to use on whether they are saving 2 to 3 years for a car, 18 to 20 years for a child's eventual college expenses, or more than 35 years for their own retirement.

Skyler: I picked Account B because it grows more than A.

Ms. Morris: What do you mean by "grows more"?

Skyler: I mean that B gets to \$5000 but A is only at, like, \$4200, so B is more.

Ms. Morris: And how long was it when you saw those amounts?

Skyler: Um, it was about 19 years.

Ms. Morris: So after 19 years, you saw much more money in Account B, $5000 versus $4200, and this led you to pick Account B? [*Skyler nods.*] Now, you said Account B grew more. Does that mean Account A will never get as high as Account B? Will Account A never reach $5000?

Skyler: No, I mean that Account B was growing faster. It got to $5000 first.

Ms. Morris: OK, I understand. We can say that Account B was growing faster than Account A because it got to $5000 first. [*pause*] And were you thinking about the college savings? Is that why you zeroed in on 19 years?

Skyler: Yeah. I was looking at 18 or 19 years.

Ms. Morris: [*Rosa raises her hand.*] Yes, Rosa?

Rosa: I don't think Account B was growing faster at the start, because I did the car savings, and Account A was faster for the first couple of years.

Ms. Morris: What did you find for car savings? Did you look at 2 1/2 years?

Rosa: Yeah, I got $2300 for Account A but only $2260 for Account B. So Account A was $40 more.

Ms. Morris: This is making me think that it's not so easy to say which account grows faster and which grows slower. It kind of depends on when we're looking at it, doesn't it? For the first few years, Account A is growing faster and has a higher value, but then by 18 or 19 years we can see that Account B is growing faster. Let's take a look at what happens way out at 35 years, and then I want us all to try to figure out exactly what is going on. When is Account A a better investment, and when is Account B a better investment? OK, so who was thinking about retirement savings?...

A content standard within the *Common Core State Standards for Mathematics* is for students to observe that "a quantity increasing exponentially eventually exceeds a quantity increasingly linearly" (CCSSI 2010, F-LE.3). By attending very carefully to students' expressions of ideas, Ms. Morris not only set the stage for this standard but also pushed students to distinguish between short-term, mid-term, and long-term growth of the linear function versus the exponential function. Following this excerpt, Ms. Morris also explored the point at which the values of the two accounts were equal. Thinking precisely about growth, however, reveals that Account B must actually start *growing* faster than Account A even before the intersection of the functions, otherwise Account B would never catch up (and then surpass) Account A.

It is important to note that Ms. Morris's interactions with Skyler at the beginning of the excerpt involved not only a push for precision but also interpersonal dynamics between Ms. Morris and Skyler. Ms. Morris felt that Skyler was confident enough in her mathematical participation to move forward productively from Ms. Morris's clarifying questions. If it had been a different student, perhaps one who was timid in sharing ideas, Ms. Morris may have chosen to pursue the attention to precision differently.

Critiquing quantifiers. Claims are the coin of the realm in mathematics, and claims come with quantifiers (e.g., for all, for some, there exists). This means that one area where students should be expected to practice attending to precision is with the quantifiers of their claims or conjectures. Tasks or teacher questions that involve always/sometimes/never distinctions are one way to promote this type of attention. At other times, it is the extent of a claim's hypothesis that must be attended to. In the excerpt below, which comes from an advanced algebra course, Ms. Hernandez prompts her class to consider a student's claim about vertical asymptotes, guiding them to refine the hypothesis rather than the conclusion of the claim.

Ms. Hernandez: I want to discuss an interesting idea that Shari had. Shari, can you repeat what you were just saying to me?

Shari: I was just saying that whenever there's a variable in the denominator, the function has a vertical asymptote.

Ms. Hernandez: I want to write Shari's claim up here on the board because it contains some important ideas for all of us to think about. We've been dealing with rational functions for quite awhile now, and Shari is saying [*writes on board*], "If a rational function has a variable in the denominator, then the function has a vertical asymptote." Did I get that right, Shari?

Shari: Yeah.

Ms. Hernandez: OK. And we've been working on finding those vertical asymptotes by checking for division by zero. But Shari is making a broader claim about the existence of the vertical asymptotes rather than saying exactly where they are. Do people agree or disagree with this claim on the board? [*pause*] Devon?

Devon: I think it's basically right, because if you have an x in the denominator then there might be a number for that x that makes it zero, which would be a vertical asymptote.

Ms. Hernandez: So I heard you say you agree with the claim, but then you said there "might" be a zero in the denominator. Is there always a zero if you have an x down there, or is it "might" or "maybe"?

Chad: I think there's always a zero, because if you have an x in the bottom then you take the bottom and set it equal to zero. When you solve that, it gives you your zero and the vertical asymptote.

Lainey: I think that sometimes there is no zero, even when you have an x in the bottom.

Ms. Hernandez: Can you say a bit more about that, Lainey?

Lainey: Well, like with x-squared plus 1. I remember that that one is never zero. So then I think if x-squared plus 1 was in the denominator, it would be OK. You wouldn't have division by zero.

$$f(x) = \frac{3x^3 - 5x - 2}{x^2 + 1}$$

Fig. 6.4. An example of a rational function

Ms. Hernandez: [*Writes an example function on the board (see fig. 6.4).*] Remember that a rational function is a function with a polynomial in the numerator and a polynomial in the denominator. So Lainey is saying that with the polynomial x-squared plus 1 in the denominator, you never have division by zero, so you don't have any vertical asymptotes. In fact, there are other polynomials that don't have any zeros, don't have any x-intercepts, right? Any parabolas that are entirely above or entirely below the x-axis or any fourth-degree or sixth-degree polynomials that are like that, too. So is it OK if I rewrite your claim, Shari?

Shari: Yeah, I get it now.

Ms. Hernandez: I'm going to write it this way [*writes on board*], "If a rational function has a *polynomial* in the denominator that has an x-intercept, then the function has a vertical asymptote." So this is our new claim. But we need to give it as much scrutiny as we did with the first claim. I'm going to give you a few minutes to look back through your work and think with each other about whether you agree or disagree with this new claim.

In this excerpt, attention to precision was integrated seamlessly with mathematical reasoning—refining a claim and reasoning about that claim are inseparable. This link results from the fact that mathematical statements are taken to be precise, meaning exactly what they say and for all cases included in the statement, with no exception. Ms. Hernandez could have taken Shari's original remark and said that it was essentially correct, linking it to the procedure for determining the location of a vertical asymptote. Instead, she led the class in the collective practice of attending to precisely what was implied by Shari's claim, which was written on the board to aid in careful consideration. Moreover, by continuing after the end of the excerpt to discuss situations in which a zero in the denominator does *not* produce a vertical asymptote (e.g., $h(x) = (x + 1) / (x^2 - 1)$ has a hole at $x = -1$ rather than a vertical asymptote), this class's attention to the precision of the claims had the potential to generate new conceptual understanding to complement the procedures for finding vertical asymptotes.

The examples above include tasks, questions, and classroom interactions that center on attention to precision. In some cases, this attention is focused on numerical calculations or measurements, but in other cases it is focused on the mathematical language being used. In all cases, the practice of attending to precision is developed over long periods of time, not in a single lesson. It may begin with thoughtful teachers modeling their own attention to precision before gradually raising expectations of the ways in which students attend to precision in their own work and the work of their classmates. These expectations should not be enforced arbitrarily with students. Instead, we should help students see the value and power in mathematical precision.

Resources

Books and Book Chapters

This book focuses on various aspects of mathematical communication in middle-grades classrooms and includes chapters written by teachers who have been thinking carefully about their discourse practices. Helping students communicate their mathematical ideas involves attending to the precision of what they are saying and writing.

- Herbel-Eisenmann, B., and M. Cirillo, eds. *Promoting Purposeful Discourse: Teacher Research in Secondary Math Classrooms.* Reston, Va: National Council of Teachers of Mathematics, 2009.

This book contains various articles collected from the NCTM practitioner journals that deal with mathematics discourse and communication.

- Elliott, P., and C. M. Garnett, eds. *Getting into the Math Conversation.* Reston, Va: National Council of Teachers of Mathematics, 2008.

This book deals with the process of teaching challenging mathematics to English language learners (ELLs) and developing their language skills. Recognizing and becoming comfortable with the precision of language is an important part of this process, and one that can benefit all students, not just ELLs.

- Ramirez, N., and S. Celedon-Pattichis. *Beyond Good Teaching: Advancing Mathematics Education for ELLs.* Reston, Va: National Council of Teachers of Mathematics, 2012.

Online Resources

Random Drawing Tool—Sampling Distribution: Larger sample sizes can increase the precision of an estimated statistic (grades 9–12).

- http://illuminations.nctm.org/ActivityDetail.aspx?ID=159

Pan Balance—Numbers: Students must precisely balance numerical expressions. This activity also foregrounds the equals sign as a relationship, not a signal to compute (grades 3–8).

- http://illuminations.nctm.org/ActivityDetail.aspx?ID=26

Computing Pi: By increasing the number of sides of the regular polygon used to approximate a circle, the area and circumference of the circle can be determined to greater degrees of precision (grades 6–12).

- http://illuminations.nctm.org/ActivityDetail.aspx?ID=161

Numbers and Language: By considering mathematical terms used in everyday situations, students can become more precise in their understanding of mathematical language (grades 3–5).

- http://illuminations.nctm.org/LessonDetail.aspx?id=U86

Journal Articles

The description of the attend-to-precision practice specifically includes elementary students giving "carefully formulated explanations to each other." In this article, a fourth-grade teacher writes about how she achieved this in her classroom.

- Kinman, R. L. "Communication Speaks." *Teaching Children Mathematics* 17, no. 1 (2010): 22–30.

The attend-to-precision practice also mentions using the equals sign "consistently and appropriately." This article presents a lesson that elementary teachers can use to address misconceptions that students may have about the equals sign.

- Mann, R. "Balancing Act: The Truth behind the Equals Sign." *Teaching Children Mathematics* 11, no. 2 (2004): 65–70.

This article uses the concept of slope as an example of moving from students' informal language to a precise mathematical definition. This mathematical progression is presented as an alternative to the more traditional mode of defining terms by rote.

- Herbel-Eisenmann, B. A. "Using Student Contributions and Multiple Representations to Develop Mathematical Language." *Mathematics Teaching in the Middle School* 8, no. 2, (2002): 100–105.

Although estimation does not lead to precise calculation results, it is valuable to understand the ways in which estimation can be useful for reasoning and sense making. This article deals with estimation strategies in the context of fraction computations.

- Johanning, D. I. "Estimation's Role in Calculations with Fractions." *Mathematics Teaching in the Middle School* 17, no. 2 (2011): 96–102.

This article contains a discussion of the role of zeros after the decimal point and their relation to precision in real-world applications.

- Drum, R. L., and W. G. Petty Jr. "2 Is Not the Same as 2.0!" *Mathematics Teaching in the Middle School* 6, no. 1 (2000): 34–40.

Within a Standards-based classroom, a teacher guided students as they moved from student-generated terminology for the *y*-intercept to a formal definition. This article also describes how this shift presented an opportunity to explicitly talk about the nature of mathematically precise definitions.

- Davis, J. "Connecting Students' Informal Language to More Formal Definitions." *Mathematics Teacher* 101, no. 6 (2008): 446–52.

The author argues that mathematics teachers often ignore significant digits and their relation to precision. He offers five useful principles that will help students attend to precision.

- Nowlin, D. "Precision: The Neglected Part of the Measurement Standard." *Mathematics Teacher* 100, no. 5 (2007): 356–62.

References

Common Core State Standards Initiative (CCSSI). *Common Core State Standards for Mathematics.* Washington, DC: National Governors Association Center for Best Practices and the Council of Chief State School Officers, 2010. http://www.corestandards.org/assets/CCSSI_Math%20Standards.pdf.

Gibbons, P. *English Learners, Academic Literacy, and Thinking: Learning in the Challenge Zone.* Portsmouth, N.H.: Heinemann, 2009.

Lakatos, I. *Proofs and Refutations: The Logic of Mathematical Discovery.* Cambridge, England: Cambridge University Press, 1976.

Moschkovich, J. "Beyond Words to Mathematical Content: Assessing English Learners in the Mathematics Classroom. In *Assessing Mathematical Proficiency*, edited by A. H. Schoenfeld (vol. 53, pp. 345–52). Cambridge, England: Cambridge University Press, 2007.

National Council of Teachers of Mathematics (NCTM). *Principles and Standards for School Mathematics.* Reston, Va: NCTM, 2000.

Schleppegrell, M. "The Linguistic Challenges of Mathematics Teaching and Learning: A Research Review." *Reading and Writing Quarterly* 23 (2007): 139–59.

Silver, E. A., L. J. Shapiro, and A. Deutsch. "Sense Making and the Solution of Division Problems Involving Remainders: An Examination of Middle School Students' Solution Processes and Their Interpretations of Solutions." *Journal for Research in Mathematics Education* 24 (1993): 117–35.

Look for and Make Use of Structure

Practice 7: Look for and make use of structure

Mathematically proficient students look closely to discern a pattern or structure. Young students, for example, might notice that three and seven more is the same amount as seven and three more, or they may sort a collection of shapes according to how many sides the shapes have. Later, students will see 7×8 equals the well-remembered $7 \times 5 + 7 \times 3$, in preparation for learning about the distributive property. In the expression $x^2 + 9x + 14$, older students can see the 14 as 2×7 and the 9 as $2 + 7$. They recognize the significance of an existing line in a geometric figure and can use the strategy of drawing an auxiliary line for solving problems. They also can step back for an overview and shift perspective. They can see complicated things, such as some algebraic expressions, as single objects or as being composed of several objects. For example, they can see $5 - 3(x - y)^2$ as 5 minus a positive number times a square and use that to realize that its value cannot be more than 5 for any real numbers x and y. (CCSSI 2010, p. 8)

Unpacking the Practice

Looking for structure, or what Cuoco, Goldenberg, and Mark (1996) call "pattern sniffing," is a fundamental mathematical habit of mind. The practice of looking for and using structure is a building block of mathematical knowledge; it gives purpose to students' work with numbers and explorations of shapes and geometric objects because the patterns and conjectures that students articulate from these explorations can be used to generate new knowledge. Practice 7 builds upon the recommendations of the National Council of Teachers of Mathematics' (NCTM 2000) *Principles and Standards for School Mathematics* in the Connections and Representation Process Standards as well as the Number and Operations, Algebra, and

Geometry Content Standards. In addition, practice 7 develops conceptual understanding, one of the strands of mathematical proficiency in *Adding It Up* (NRC 2001).

Connections Standard

It has been said that mathematics is the science of patterns (Devlin 1996), and pattern-recognition and pattern-generalizing activities have become more prominent in K–12 mathematics since the publication of NCTM's (2000) *Principles and Standards for School Mathematics.* Practice 7 emphasizes the importance of mathematical structure as a common root of patterns and generalizations, which means that students' abilities to recognize and use structure will help them discover new mathematical knowledge. For example, students might discover the pattern in products of two even numbers—that the result is always even—and then link this discovery to the structure of even numbers. Because all even numbers share a similar structure, this observation may yield a statement that applies to all products of even numbers. When proven mathematically, this result allows students to use this knowledge to justify why products of three even numbers are even. Recognizing relationships that apply to a large set of cases helps students develop a conceptual understanding of how mathematics is a coherent, highly structured system of ideas and how expressing and using this structure makes mathematics much easier to understand.

Practice 7 also emphasizes the importance of students recognizing relationships between previously determined results and new ideas. Seeing structure is about making connections at a conceptual, abstract level among different mathematical forms and ideas, which in turn condenses many seemingly different mathematical objects into a much smaller, related set. For instance, when students recognize the structure of the commutative property, their "learning of basic addition combinations is reduced by half" (NRC 2001, p. 120). The Connections Standard for kindergarten through grade 12 in *Principles and Standards for School Mathematics* (NCTM 2000) says that "as students progress through their school mathematics experience, their ability to see the same mathematical structure in seemingly different settings should increase" (p. 65).

Being able to *see* structure often means a student needs to "step back for an overview and shift perspective" (CCSSI 2010, p. 8). Sfard (1994) famously asked the community of mathematics educators: "When you look at an algebraic expression such as, say, $3(x + 5) + 1$, what do you see? It depends." Sfard (1994) defined *reification* as what happens when one shifts from seeing mathematical objects as a series of operations to conceptualizing and operating on them as a whole structure. In the case of $3(x + 5) + 1$, one might see this expression as a series of actions to take with an unknown number (an operational view) or as a mathematical object with a particular structure. The ability to shift perspective becomes more relevant when the expression is $(x + 5)^2 + 2(x + 5) - 1$. By seeing the expression as a whole object, students can recognize that it follows the same structure as the factorable $x^2 + 2x + 1$.

The NCTM (2000) Connections Standard discusses how helping students recognize the connections among mathematical ideas, such as seeing the structure of $(x + 5)^2 + 2(x + 5) + 1$ and connecting it to their work with factorable trinomials, involves planning lessons with tasks that afford students the opportunity to see such connections. For instance, helping students develop an ability to reify expressions and equations in algebra may require shifting standard, operational approaches to the algebra curriculum to more functional, object-based approaches (Sfard and Linchevski 1994). Students tend to learn about expressions—writing them and manipulating them—before learning to write and solve equations. The emphasis in students' early algebra work is on operating with terms, not on considering the relationships among variables. Sfard and Linchevski suggest that a functions-based approach to learning algebra, where students come to understand algebraic letters as variables instead of simply unknowns and where relationships between variables are emphasized, can help students gain fluency in recognizing equivalence by taking multiple perspectives on mathematical objects.

Representation Standard

As stated in the Representation Standard, developing the ability to create and use representations to organize, record, and communicate mathematical ideas can help make students' mathematical ideas "available for reflection" (NCTM 2000, p. 67). For instance, being able to express the distributive property using algebraic symbols ($a(b + c) = ab + ac$, where a, b, and c are real numbers) allows a student to not only recognize the distributive property when multiplying binomials (e.g., $(x + 3)(x + 5) = x \cdot x + 5x + 3x + 3 \cdot 5$) but also see how decomposing numbers such as 26 into $2(12 + 1)$ is a situation where the distributive property applies. One way for students to understand the importance of algebraic notation is to recognize how symbols help represent relationships in general ways that can be applied to new situations. The Algebra Standard states that "instructional programs from prekindergarten through grade 12 should enable all students to...represent and analyze mathematics situations and structures using algebraic symbols" (NCTM 2000, p. 37). For example, expressing an explicit rule for the pattern 1, 4, 7, 10,... that can provide any term in the sequence, whether it is the 100th or 10,000th term, depends upon the ability to write the relationship between the term number and the value in the sequence using algebraic symbols, such as $y = 3x + 1$, where x and y are whole numbers. This equation reveals the structural relationship embedded in the pattern as opposed to the recursive relationship that students often initially define. Although algebraic symbols are typically letters, they need not be as long as one clearly defines which symbols represent the dependent and independent variables.

The Role of Assumption in Mathematical Structures

One part of developing competency in doing mathematics involves being aware of and understanding when to apply *assumptions* (mathematical statements either known to be, proven to be, or assumed to be true). One example of this is drawing an auxiliary line to a figure while doing a construction or generating a proof in geometry. Consider the claim that triangles with the same area are not necessarily congruent (Chazan 1990). In this proof, $\triangle ABC$ is constructed with D as the midpoint of $\overline{AB}$ and $\triangle ABC$ scalene (see fig. 7.1).

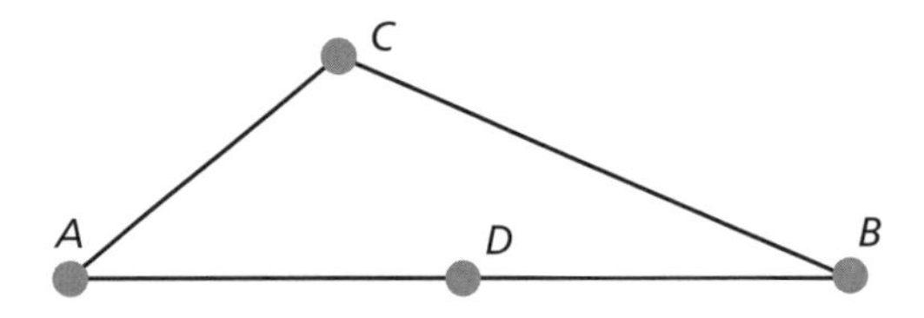

Fig. 7.1. Scalene triangle *ABC*

Median $\overline{CD}$ divides $\triangle ABC$ into two triangles of equal area (because they have equal bases and equal heights). Thus, drawing the medians of $\triangle ACD$ and $\triangle DCB$ from vertex *D* will yield four smaller triangles, all of equal area (see fig. 7.2).

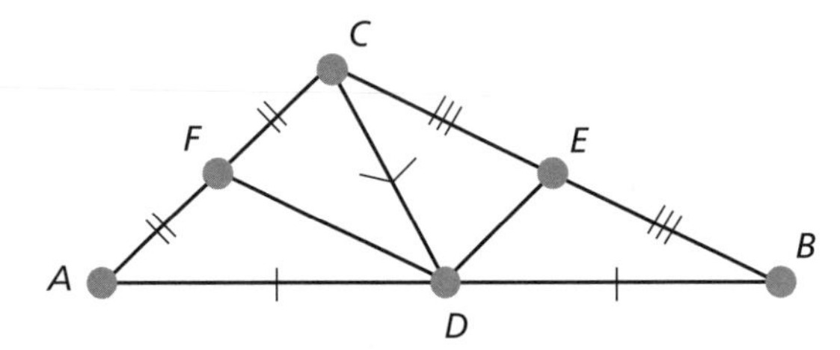

Fig. 7.2. Scalene triangle *ABC* with medians

However, the four smaller triangles are not necessarily congruent to one another. Because, by definition of scalene, the measures of angles A, B, and C cannot be equal, we do not have given conditions such that congruence can be proved. Thus, by drawing medians from vertex *D* as auxiliary lines, we can provide a counterexample to the claim that triangles with the same area are necessarily congruent.

Another example of how recognizing when a mathematical object or statement can be used or assumed to be true is in the case of elementary arithmetic properties. The basic multiplicative identity ($a \times 1 = a$, for all $a \in \mathbb{R}$) is useful throughout algebra, precalculus, and beyond as a technique for solving equations and integration. Consider the problem $\int x\sqrt{x^2 - 1}\,dx$. An elegant technique for integrating this function is to do a *u*-substitution, where $u(x) = x^2 - 1$, and so $\frac{du}{dx} = 2x$. However, the function being integrated is stated as $x \cdot u \cdot dx$, not $u \cdot \frac{du}{dx}$.
This is where the multiplicative identity can help; by multiplying by a factor of 1, specifically $2/2$, then the constant $1/2$ can be moved outside the integral and now the problem becomes $\frac{1}{2}\int 2x\sqrt{x^2 - 1}\,dx$ or $\frac{1}{2}\int \sqrt{u}\,du$. Now, students can complete this much simpler integration and substitute $u(x) = x^2 - 1$ afterwards. Looking back, the strategy involved recognizing the structural relationship between the expression inside the radical and the *x* outside the radical, namely, *x* is half the derivate of $x^2 - 1$.

Classroom Examples

Inside classrooms, it can be challenging to engage students in this practice of looking for and making use of structure when textbooks often produce rules describing properties such as the commutative property ($a + b = b + a$, where $a, b \in \mathbb{R}$) and the distributive property ($a(b + c) = ab + ac$, where a, b, and $c \in \mathbb{R}$), and place these rules in attention-grabbing boxes labeled "Property" or "Theorem." In so doing, the textbook reduces the cognitive demand (Stein, Grover, and Henningsen 1996) on students to recognize these properties as they emerge in their arithmetic work. In this section, we provide some vignettes developed from actual lessons in elementary and secondary classrooms to illustrate how classroom practice can support students' engagement with this mathematical practice.

Elementary Grades Vignette: Addition and Subtraction with Ms. Wilcox

The following vignette is inspired by a lesson taught in Tracy Lewis's second-grade class at Anna Yates Elementary School in Emeryville, California (video of this lesson, as well as teacher commentary and work samples can be found at http://www.insidemathematics.org). In this fictional lesson, students are learning to solve problems involving addition and subtraction in context and are taught by a fictional teacher named Ms. Wilcox. She gave three tasks to students at the beginning of the lesson:

1. The apple farm is 92 miles from the school. They have traveled 58 miles so far. How many more miles do they have to go? Show how you know your answer is correct.
2. Molly wants to buy an apple pen. She sees a red apple pen that costs 48 cents. She sees a sparkle apple pen that costs 65 cents. How much more does the sparkle apple pen cost? Show how you know your answer is correct.
3. Time to leave! 36 students got on the bus. They waited until all 63 students were there. How many students were late to the bus? Show how you know your answer is correct.

The class has just finished discussing students' work on problem 1. After discussing three solutions, the class decided that the following two solutions were correct:

Juvena's response: I know that 50 plus 42 is 92. They have traveled 58, not 50 miles. So, they need to go 34 miles, which I found by taking 8 away from 42.

Shaun's response: I drew a picture [*see fig. 7.3*]. In my picture, each mark stands for 10 miles. They've gone 58 miles. So, they need to go 2 more miles to reach 60. Then, they only have 32 miles to go to reach the apple farm. So, they still have to go a total of $2 + 32 = 34$ miles.

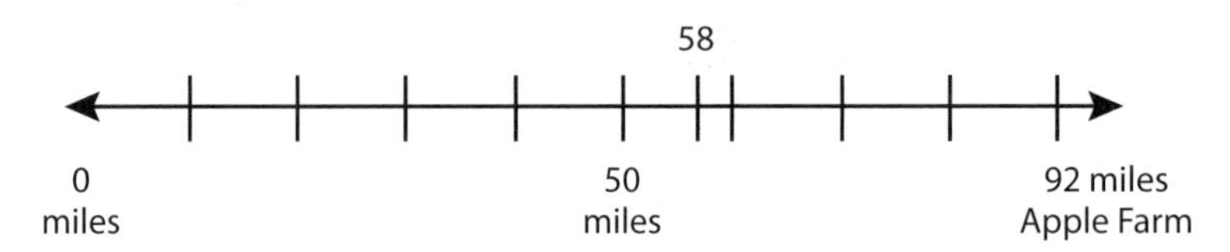

Fig. 7.3. Shaun's picture

[Before beginning work on the second problem, Ms. Wilcox engages her students in considering the different approaches used by Juvena and Shaun.]

Ms. Wilcox: So we agree that both Juvena and Shaun got the correct answer and used strategies that were correct. Now, I'd like someone to tell us: what are the differences between Juvena's work and Shaun's work? [*Pauses until several hands are raised*] Sheila?

Sheila: Shaun drew a picture, but Juvena just kind of thought about it and did the math.

Ms. Wilcox: These observations are true, Sheila, but I'd like for us to think more deeply about the problem. What is different about the mathematics that Shaun and Juvena used to solve the problem? [*Pauses for about 20 seconds—no hands are raised*] Take a look at the first step they did. What is different about the first step? Yes, Shaun?

Shaun: Juvena started by figuring out how many miles they had left to go, but she used 50 instead of 58 at first. I started by taking how many miles they had traveled already, 58, and then adding 2 to get 60. Then, I figured out how many more miles they had left to go.

Ms. Wilcox: Okay, thank you Shaun. I'm going to write on the board the computations so everyone can see what you are referring to. So, Juvena started by subtracting 8 from 58 to get 50. [*Writes 58 – 8 = 50*] Then, she figured out how many more miles they needed to go to get to the apple farm. [*Writes 50 + 42 = 92*] But, because they had actually traveled 58, not 50 miles, she had to subtract 8 from 42 to get the correct answer. [*Writes 42 – 8 = 34*] Shaun started out differently. He took the 58 miles the group had traveled and then added 2 to get to 60 miles. [*Writes 58 + 2 = 60*] Then, Shaun added to figure out how many more miles they would need to travel to make it to the apple farm. [*Writes 60 + 32 = 92*] Because he had already added in 2 miles to get to 60, he added 2 more miles to 32 to get the correct answer of 34 miles. Okay, what I would like you to do now is look at problem 2. [*Ms. Wilcox reads the problem aloud.*] I want you to first work in your small groups to solve problem 2 using the same method as Shaun.

[*The class takes about five minutes to work on problem 2. Ms. Wilcox spends time listening to each group work and selects the members of Molly's group to share their work on problem 2.*]

Ms. Wilcox: Let's talk about problem 2. I've asked Molly to share her group's work on problem 2. As Molly shares her work, I want everyone to listen carefully to see whether or not Molly's group has applied the same method that Shaun used.

Molly: So, the red apple pen costs 48 cents, so we said that if we add 2 cents we get to 50 cents. The sparkle apple pen costs 65 cents, and 50 cents plus 15 more cents is 65 cents. So, since we had already added in 2 cents to get to 60 cents, the sparkle apple pen actually costs 2 plus 15 cents, which is 17 cents more.

Ms. Wilcox: Does anyone have any questions for Molly? Did her group use Shaun's method? How do you know? Maryam?

Maryam: Molly's group did the same thing as Shaun because they added 2. Then, they figured out how many more cents they would need to buy the sparkle apple pen. That's like Shaun figuring out how many more miles they would need to go to get to the apple farm.

Ms. Wilcox: OK, so Maryam's noticing that both Molly's group and Shaun added 2 to the starting amount. Can you explain why they added 2? Juvena?

Juvena: They are trying to get to an even amount.

Ms. Wilcox: What do you mean by even, Juvena?

Juvena: Like 10s—Shaun added 2 to get to 60, and Molly's group added 2 to get to 50. When I did the first problem, I took away 8 to get to 50.

Ms. Wilcox: That's an important point. It's not just any even number, but it's a multiple of 10. By using amounts that are multiples of 10, it makes it easier to add or subtract. So, for the final problem, I'd like you to use Shaun's method again to solve this problem in your small groups. I will walk around to listen as you solve the problem, and once everyone has an answer we will talk about how you applied Shaun's method.

This vignette illustrates a class reasoning about situations involving subtraction. The method featured in the vignette, Shaun's method, is a method of counting on to find the difference. In this case, students must add on from the lesser amount to arrive at a multiple of 10. Then, students add on from the multiple of 10 to the larger amount to find the rest of the difference between the two amounts. The important part of the method's structure is to use multiples of 10 to make the computation easier.

There are aspects of the tasks, and how Ms. Wilcox facilitates discussion of the tasks, that support students in seeing and making use of structure. First, the tasks are designed so that the first two problems feature quantities that are two digits away from a multiple of 10. The design of the tasks makes it easier for students to recognize the proximity of that quantity to a multiple of 10. The third task also involves a difference of two quantities, but the lesser quantity is six digits away from a multiple of 10. With this difference between problem 3 and problems 1 and 2, the teacher can assess whether students are recognizing the important feature of the underlying structure of Shaun's method (computing with a multiple of 10) and not just using surface-level similarities (you always add 2 in the first step).

In the vignette, Ms. Wilcox's questioning strategy also supports students in seeing structure among the problems and Shaun's method. After both Juvena and Shaun present their work, she has the class consider what is different between the two methods as a way of highlighting the inherent structure of each solution strategy. When Sheila responds that one difference is that Shaun "drew a picture" and Juvena "did the math," Ms. Wilcox takes the opportunity to point out that students should notice differences in the mathematics of the two approaches.

After Shaun summarizes the differences in how he and Juvena started their solutions to problem 1, Ms. Wilcox provides a summary of the strategies on the board. This summary is a launch to the next key move Ms. Wilcox makes to support students in seeing structure: asking students to solve a new problem, with a different context, using Shaun's method. This activity forces students to consider the general features of the method and determine how to apply those steps in a new situation. Also, Ms. Wilcox continues to draw students' attention to articulating the general features of Shaun's method when having Molly share her group's work on problem 2 with the class. Instead of asking the class to determine whether Molly's group obtained the correct answer, Ms. Wilcox asks the class to consider whether Molly's group applied the same method as Shaun.

As discussed earlier, the differences between problems 2 and 3 are significant in terms of assessing whether students are recognizing the structural features of Shaun's method and not just attending to similarities in the amounts that are being added or subtracted across each problem. Ms. Wilcox explicitly asks the class if Molly's group applied the same method as Shaun. Maryam's response indicates that she thinks they did because "they both added 2," and Ms. Wilcox again seizes an opportunity to address students' misconceptions by asking "Can you explain why they added 2?" This takes the focus away from the quantity being added and forces students to consider how what is being added is part of a solution strategy. The question also allows the major structural feature of the strategy to be highlighted—"getting to multiples of 10"—before students work on problem 3.

This vignette illustrates that engaging students in looking for and using structure involves attention to the structure inherent in solution methods and how differences between tasks in a problem set can highlight the overarching mathematical goal for the lesson. The tasks cannot stand alone; the questions that the teacher poses and how the teacher directs students' attention to features of the tasks and students' work on the tasks is paramount to helping students see and use structure in mathematics. This kind of mathematical work is not, however, unique to elementary mathematics or to particular tasks. In the section that follows, a vignette from a high school algebra class illustrates more ways that students can learn to recognize and use structure in the school mathematics curriculum.

Secondary Grades Vignette: Finding the Missing Coefficient

This secondary vignette, from a high school algebra classroom, highlights a different part of practice 7—namely, being able reflect on a problem and shift perspective to find a solution. Sfard (1994) discusses the importance of students' abilities to *reify*, in essence to consider how mathematical operations can be understood as both process and product, in the context of algebra. A basic example of this occurs when recognizing that the expression $5x + 3$ can be seen as distinct operations "5 copies of x added to 3" as well as a product "the quantity obtained by computing $5x + 3$." In the vignette that follows, we present a lesson that involves reification in representing linear functions. Inspired by Knuth (2000a, 2000b), this vignette features a whole-class discussion of how to find the missing coefficient of x for the equation $?x + 2y = -4$. Students were provided with a graph of the equation (see fig. 7.4) and were given time in class to discuss the problem in small groups.

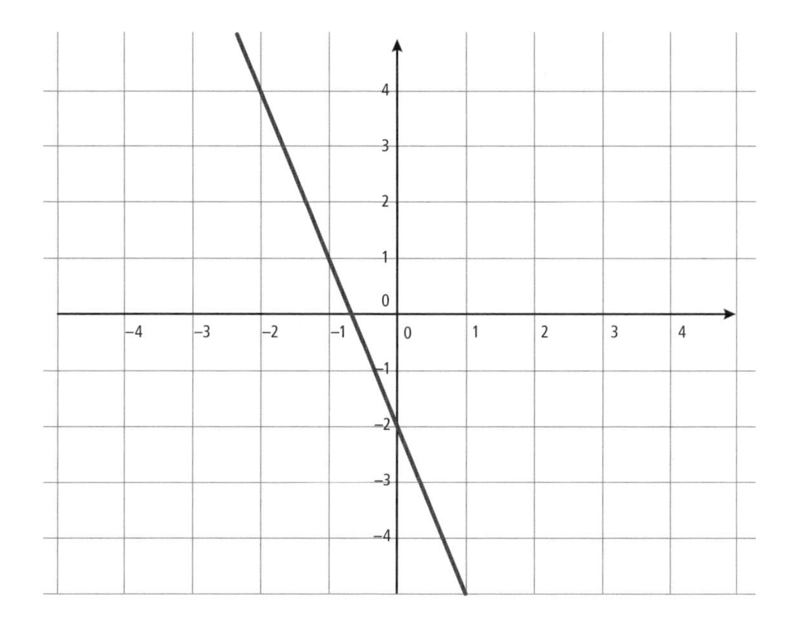

Fig. 7.4. Graph of $?x + 2y = -4$.

As the teacher, Mr. Abel, walked around to different groups, he noticed a couple of strategies the groups were using to solve the problem. One strategy was to rewrite the equation into slope-intercept form, then find the slope of the line from the graph and set that value equal to the expression for m in the slope-intercept form of the equation. Another strategy was to pick a point from the graph, substitute the x and y values from that point into the equation and solve the equation for the missing value. Some groups were stuck; they didn't think there would be a way to figure out the missing value. Mr. Abel decides to select one group to share the slope-intercept strategy and another group to share the substitution strategy. The vignette begins as Mr. Abel is encouraging students to step back and think about the similarities and differences between the two strategies:

Mr. Abel: So, Sara's group showed us how you could start by rewriting the equation into slope-intercept form, then setting the m value equal to the slope of the line from the graph and solving for the question mark. Her group found that the question mark was equal to 6. Todd's group approached the problem differently. Can anyone describe, in their own words, the strategy that Todd's group used? Yes, Marti?

Marti: They picked the point (–2, 4) from the line. Then, they let x be –2 and y be 4 in the equation. They solved the equation for the question mark and got the same answer.

Mr. Abel: Todd, do you agree? Is that what your group did?

Todd: Yeah.

Mr. Abel: OK, so the question is, why could Todd's group do that? Danny?

Danny: So, we tried to do something like that. We picked the point (0, –2) and tried to substitute it in but we got a weird answer: 0 = 0. It didn't tell us anything.

Mr. Abel: Interesting. Mark, you were in Danny's group. Why did your group decide that they could pick the point (0, –2) and use that to solve for the missing value?

Mark: I think it was Cassie who suggested it, but we thought that we remembered that any point on the line can be an x and y value in the equation. If that is the graph of this equation, then any point on the line will work. We picked (0, –2) because it was clear what the values were.

Mr. Abel: So, your group thought that any point on the line could be an x and y value in the equation. I'd like to get back to Danny's original question about why picking (0, –2) didn't help them. But, now, I would like to hear from someone else as to whether Danny's group is right—is it always the case that any point on the line can be substituted for the x and y values in the equation? What do you think, Ariana?

Ariana: Well, if you think about it, when we graph equations, we usually generate a table of values by plugging in some values we pick for y and then solving for x. So, we plot those points and construct the line. What Danny's group did is like doing this in reverse—you can take the x and y values from the graph and use them to find the missing value in the equation.

Mr. Abel: OK, so Sara, Todd, and Danny's groups all used the graph in some way to help them find the missing value, 6. Can someone tell us the differences between how Sara's group and how Todd and Danny's groups used the graph in their solution? Yes, Kristina?

Kristina: I think Sara's group used the graph to find the slope, then they solved for the question mark. Todd and Danny's group actually used the graph to find possible values for x and y, then solved for the question mark. Both methods work, I guess.

Mr. Abel: You're right, Kristina, both methods obtained the same answer. Is there any reason that you might want to use one method over the other?

Kristina: I don't know. Maybe sometimes it's harder to find the slope, like if the line doesn't cross the grid lines at an exact point.

Mr. Abel: Ivan, I see your hand raised. What do you think?

Ivan: I think it's easier to go with what Todd's group and what Danny's group did, because you only have to solve once. You just plug in the values for x and y and solve for the question mark, but in Sara's group's way you have to put it into slope-intercept form first, then find the slope so you can solve for the missing value. It seems like it takes more steps to do it the way Sara's group did it.

Mr. Abel: Would others like to weigh in on this? Yes, Lexi, we haven't heard from your group yet.

Lexi: So, we were totally stuck on this one. We didn't even realize that the points on the line could be x and y values in the equation.

Mr. Abel: Do you think that is the case now? What convinced you?

Lexi: Yes, I think so. It was how Ariana explained it—that made sense because of how we find points in our table before we draw a line. We just didn't make the connection.

Knuth (2000a), citing Moschkovich, Schoenfeld, and Arcavi (1993), described the "connection" that Lexi mentions as the *Cartesian connection*. Simply stated, the Cartesian connection is the relationship that a point is on the graph of a line if and only if its coordinates are solutions of the linear equation represented by the line. In the vignette, students who were able to recognize and apply the Cartesian connection, like the students in Todd's group, were able to quickly find a solution for the missing value. However, as Knuth (2000a) points out, students' typical experiences in mathematics classrooms and the types of tasks in high school algebra texts overemphasize unidirectional work with linear representations; students are typically given the linear equation to graph or asked to write an equation from a graph, and only one representation is provided for tasks involving solving a linear equation. As a result, students may lose sight of the connection between the representation of an equation and the solution(s) of the equation itself.

Applying the Cartesian connection or recognizing when an auxiliary line can be useful in a geometry proof both involve flexible thinking and a conceptual understanding necessary to shift perspective while working on a problem. Developing conceptual understanding is an essential component to engaging students in the practice of recognizing and making use of structure. Students may be able to carry out procedures, such as identifying y-intercepts from the graph of a line or finding the x-intercepts of a quadratic equation, but they are unable to perform those same tasks using multiple representations of an equation without conceptual understanding of the relationship between representations of equations. The ability to make use of structure can enhance performance on problem-solving tasks; especially in complex, real-world situations, students may need to think creatively and recognize structure that may be hidden within context to model the situation mathematically. As we have mentioned throughout the chapters, there is a clear overlap between students' abilities to engage in practice 7 and employing other mathematical practices. All the mathematical practices require a rich understanding of mathematical ideas and demand more than just procedural fluency.

Resources

Tasks

The Horseshoes in Flight task (available at http://www.nctm.org, see link below) is designed to have students "analyze the structure of algebraic expressions and a graph to determine what information each expression readily contributes about the flight of a horseshoe. The task also illustrates a step in the mathematical modeling process that involves interpreting mathematical results in a real-world context."

- Horseshoes in Flight: http://www.nctm.org/uploadedFiles/Journals_and_Books/Books/FHSM/RSM-Task/Horseshoes.pdf#search=%22horseshoes in flight%22

Bank Shot: This task (available at http://www.nctm.org, see link below) "is intended to involve multiple geometric perspectives and would be appropriate for students with an understanding of similar triangles, rigid motions (reflections), and equations for lines and is designed to develop students' understanding of these concepts."

- Bank Shot: http://www.nctm.org/uploadedFiles/Journals_and_Books/Books/FHSM/RSM-Task/Bank%20Shot.pdf#search=%22bank shot%22

Websites

This site provides video examples of classroom lessons across grade levels that showcase students applying practice 7. Student work samples and teacher commentary are provided with each video.

- InsideMathematics: www.insidemathematics.org/index.php/standard-7

This site provides access to resources that can be used to determine alignment between curricula and the Standards for Mathematical Practice.

- Common Core Curriculum Analysis Tools: http://www.mathedleadership.org/ccss/materials.html

References

Chazan, D. "Students' Microcomputer-Aided Explorations in Geometry." *Mathematics Teacher* 83, no. 8 (1990): 628–35.

Common Core State Standards Initiative (CCSSI). *Common Core State Standards for Mathematics.* Washington, D.C.: National Governors Association Center for Best Practices and the Council of Chief State School Officers, 2010. http://www.corestandards.org.

Cuoco, A., E. P. Goldenberg, and J. Mark. "Habits of Mind: An Organizing Principle for Mathematics Curricula." *Journal for Mathematical Behavior* 15 (1996): 375–402.

Devlin, K. *Mathematics: The Science of Patterns: The Search for Order in Life, Mind and the Universe.* New York: Holt Paperbacks, 1996.

Knuth, E. "Student Understanding of the Cartesian Connection: An Exploratory Study." *Journal for Research in Mathematics Education* 31, no. 4 (2000a): 500–507.

Knuth, E. "Understanding Connections between Equations and Graphs." *Mathematics Teacher* 93, no. 1 (2000b): 48–53.

Moschkovich, J., A. Schoenfeld, and A. H. Arcavi. "Aspects of Understanding: On Multiple Perspectives and Representations of Linear Relations and Connections among Them." In *Integrating Research on the Graphical Representation of Functions*, edited by T. A. Romberg, E. Fennema, and T. P. Carpenter (pp. 69–100). Hillsdale, NJ: Erlbaum. (1993).

National Council of Teachers of Mathematics (NCTM). *Principles and Standards for School Mathematics.* Reston, Va: NCTM, 2000.

National Research Council (NRC). *Adding It Up: Helping Children Learn Mathematics.* Edited by J. Kilpatrick, J. Swafford, and B. Findell. Washington, D.C.: National Academies Press, 2001.

Sfard, A. "Reification as a Birth of a Metaphor." *For the Learning of Mathematics* 14, no. 1 (1994): 44–55.

Sfard, A., and L. Linchevski. "The Gains and the Pitfalls of Reification: The Case of Algebra." *Educational Studies in Mathematics* 26 (1994): 191–228.

Stein, M. K., B. Grover, and M. Henningsen. "Building Student Capacity for Mathematical Thinking and Reasoning: An Analysis of Mathematical Tasks Used in Reform Classrooms." *American Educational Research Journal* 33 (1996): 455–88.

Look for and Express Regularity in Repeated Reasoning

Practice 8: Look for and express regularity in repeated reasoning

Mathematically proficient students notice if calculations are repeated, and look both for general methods and for shortcuts. Upper elementary students might notice when dividing 25 by 11 that they are repeating the same calculations over and over again, and conclude they have a repeating decimal. By paying attention to the calculation of slope as they repeatedly check whether points are on the line through (1, 2) with slope 3, middle school students might abstract the equation $(y - 2)/(x - 1) = 3$. Noticing the regularity in the way terms cancel when expanding $(x - 1)(x + 1)$, $(x - 1)(x^2 + x + 1)$, and $(x - 1)(x^3 + x^2 + x + 1)$ might lead them to the general formula for the sum of a geometric series. As they work to solve a problem, mathematically proficient students maintain oversight of the process, while attending to the details. They continually evaluate the reasonableness of their intermediate results. (CCSSI 2010, p. 8)

Unpacking the Practice

Mathematics educators for many years have recognized the importance of students developing both the ability to perform mathematical procedures as well as a conceptual understanding of the mathematical ideas they are working with. The National Council of Teachers of Mathematics (NCTM), in articulating the Learning Principle in the *Principles and Standards for School Mathematics* (2000), made it clear that procedures and concepts could be and should be mutually supportive for students. Similarly, *Adding It Up* (NRC 2001) tied "procedural fluency" with "conceptual understanding" as two of the five strands of mathematical proficiency. The *Common Core State Standards for Mathematics* (CCSSM) echoes these positions, stating up front that "mathematical understanding and procedural skill are equally important" (p. 4). It is with this in mind that we can interpret the current practice—look for and express regularity in repeated reasoning—as providing important insights into the relationship between procedural and conceptual work in mathematics classrooms.

In particular, this practice illuminates the types of thinking that students can engage in as they work through procedures. Each of the examples presented in the description of the practice shows the potential of a fairly procedural task leading to a more general conceptual observation: an enactment of the long division algorithm can lead to a realization about repeating decimals; the calculation of slopes between various points can lead to an understanding of linear equations; the multiplication of multinomials can build intuition about the sum of geometric series. These examples suggest that conceptual understanding is a valuable outcome of procedural work and that the practice of looking for and expressing regularities can be a vehicle through which students build concepts and make sense of their procedural activities.

In addition to underscoring the value of conceptual understanding, the description of this practice also lays out an important role for students' engagement in procedures because such activity can become a site for uncovering profound mathematical relationships. In a sense, looking for and expressing regularity in repeated reasoning is a description of what mathematics students should be doing *as* they compute, calculate, manipulate, or use an algorithm. They should not be satisfied with merely executing the procedure accurately or arriving at the correct answer—after all, there are technological tools that can do this faster and more accurately than a student could ever hope to. Instead, as students work, this practice suggests that they should look for patterns, consider generalities and limitations, and make connections across past and present bouts of reasoning. These are processes beyond the capabilities of even our most advanced computer software that should be encouraged among all of our mathematics students.

The fact that this mathematical practice begins with the act of *looking for* regularities highlights the part played by students' disposition with respect to this practice. If students do not expect there to be patterns or regularities in their mathematical work—that is, if they do not expect mathematics to make sense—then it is unlikely that they will make it a common part of their mathematical activity to look for and express regularity in their repeated reasoning. Luckily, human beings seem generally predisposed to notice patterns in their experience, and as mathematics educators we should strive to cultivate that predisposition into a mathematically "productive disposition," as it is referred to in *Adding It Up* (NRC 2001).

The next two sections look more closely at issues of generalization and categorization, two important types of regularity that can be gleaned from repeated reasoning, through the lens of the Reasoning and Proof and Connections Process Standards from *Principles and Standards for School Mathematics* (NCTM 2000). Then the relationships between the practice of looking for and expressing regularity in repeated reasoning and some of the other Standards for Mathematical Practice are explored before a presentation of classroom excerpts and classroom tasks that exemplify this practice. The chapter concludes with a set of resources that can spur further consideration of these issues.

Reasoning and Proof Standard

One of the foundational acts within the process of reasoning and proof is generalizing, which produces claims that can be explored, refined, justified, and proved. Generalizing—the act of identifying commonalities, extending reasoning, or deriving broad claims from particular instances (Ellis 2011)—is also one of the most prevalent ways that the practice of looking for and expressing regularity in repeated reasoning shows up in mathematics classrooms. It not only is at the core of the discipline of mathematics, as mathematicians spend a great deal of their energy seeking, articulating, and proving generalizations, but is also receiving increased attention in school mathematics because it can be used to promote algebraic understanding. In fact, NCTM has marked generalization as an "essential understanding" for the upper-elementary grades because of its potential as the basis for algebraic thinking (Blanton et al. 2011).

Principles and Standards for School Mathematics (NCTM 2000) identifies pattern recognition as a key standard for elementary students (pp. 74, 82). Young human beings, and older human beings, for that matter, tend to be naturally inclined toward noticing regularity and commonalities across their experience. This is how we are, in some ways, hardwired to make sense of our world. The danger, however, is that students may come to turn off this pattern-recognition tendency in the context of mathematics classrooms. Instead, students may wait to be told by a teacher or textbook what they are supposed to notice and how they are supposed to proceed or may try to recall a previously learned procedure rather than searching for a pattern or mathematical relationship. Students may come to view mathematics as a domain that does not make sense and so they do not actively look for patterns to generalize.

The inclusion of this practice in the *Common Core State Standards*, therefore, can be taken as a signal to engage students in active sense making, to make them aware of the potential for generalizations as they work, and to encourage the discovery and expression of those generalizations. Thus, as mentioned above, an important foundation for this practice is students' disposition toward seeking patterns, which hinges on their recognition that mathematics is something that makes sense and a place where general relationships can be found.

Often, however, it is not generalities that receive explicit attention in mathematics classes but particular mathematical objects—specific numbers or operations, specific geometric objects, or story problems with specific conditions. So even if students have been told that general relationships often hold in mathematics, and that these relationships can be used to make sense of mathematical ideas, the students may still end up focusing the vast majority of their attention on particulars because this is what they are working on from moment to moment. NCTM in the *Principles and Standards* emphasizes the need to "move from considering *individual* mathematical objects—this triangle, this number, this data point—to thinking about *classes* of objects—all triangles, all numbers that are multiples of 4, a whole set of data" (2000, p. 188; italics in original). One way this shift to the general is promoted in classrooms is through the use of generalization tasks. For example, think of the proverbial patterning task that first asks students to identify a relationship for small numbers, then extends it to a larger, unwieldy number, and finally extends it to *any* number. Another way the shift from a focus on individual objects to classes of objects can be promoted is through teacher questions such as, "Do you think that will always work?" One of the strengths of such questions is that they are versatile. Whereas the previous example of the patterning task may only be appropriate in certain situations and is usually required to be built into the written task itself, teacher questions that promote general thinking can be used in almost any situation and need not be written into the task beforehand—they can arise in the moment of teaching.

Once students are looking for regularity in their mathematical work and have opportunities to attend to those regularities, it then becomes important for teachers to work with students as they find ways to *express* those regularities or generalizations, which links to the Communication Process Standard from the *Principles and Standards for School Mathematics.* At first, these expressions may be verbal descriptions that are informal in nature but nevertheless bring the regularity out into the open for discussion and reflection. With a teacher's guidance, students can then work to collectively inspect the regularity and canonize it in more formal language or symbols, if appropriate (e.g., shifting from numerical work to the use of algebraic notation). Overall, this process can often lead to generalizations that are new pieces of mathematical knowledge for the classroom community. Additionally, the work of expressing generalizations can lead to rich opportunities to engage in other key mathematical practices, as is discussed in a later section of this chapter.

Finally, the practice of looking for and expressing regularity in repeated reasoning can move beyond generalizing mathematical relationships to generalizing arguments. For example, a student may argue that the product of 4 and 3 is even because it can be thought of as a sum of even numbers (i.e., $4 + 4 + 4$). Another student may argue similarly that 12×7 is even because it is also equivalent to a sum of even numbers. One way to move this line of reasoning forward is to generalize the phenomenon in question, that is, by claiming that the product of an even number and an odd number is even (or, more generally yet, the product of an even number and any integer is even). Another way forward, however, would be to focus on the generality of the argument itself. There is nothing special about the numbers 4 and 3 or the numbers 12 and 7 that makes the logic hold; the only requirement is that one of the factors be even. In this way, the generality of the students' argument is a regularity in their repeated reasoning that can lead to a discussion of the fact that the argument, when viewed from the proper perspective, actually establishes the truth of infinitely many particular statements. This approach not only opens the door for construction of a viable argument (see practice 3) but also eventually leads to the general claim that the product of an even number and an odd number is even, with the proof of this claim already in hand.

Connections Standard

The Connections Process Standard involves connections between mathematics and other subject areas, between different domains of mathematics, and between different ideas within the same domain of mathematics. With respect to practice 8, regularity in students' own reasoning stems from connections across their mathematical work. Being aware of this regularity, then, may allow students to build mathematical connections and also become aware of the patterns in their own forms of reasoning. For example, at the high school level, when formal mathematical proofs are on the table, students may notice patterns in their proof techniques. They may observe that many of their proofs establish the congruence of line segments by first establishing the congruence of corresponding triangles. Students engaging in this mathematical practice over time may develop categories of mathematical arguments or subarguments and become more sophisticated consumers and producers of mathematical justifications.

Another example of this aspect of the practice is students attending to similarities (and differences) between their reasoning at a given moment and the reasoning they have engaged in previously. Such attention can result in discussions of patterns across problems and the identification of problem types. Research has shown that attention to the structure of mathematical problems is beneficial both for teachers and for students. For example, at the elementary level, teachers are better able to guide their students through learning the basic numerical operations when the teachers are aware of the different structures underlying addition and subtraction problems—such as joining/separating, part–part–whole, or comparing—and when they are aware of the different strategies that students are likely to use in those situations (Carpenter et al. 1989). At the middle school level, researchers (Jitendra et al. 2009) have also found that prompting students to consider the structure of ratio and proportion problems was one factor that led to improved problem-solving performance.

Such categories of problems are not unique to addition and subtraction or ratio and proportion. As teachers are well aware, problem types are common throughout the school mathematics curriculum. As stated in *Principles and Standards* (NCTM 2000), "much of school mathematics can be seen as the codification of answers to sets of interesting problems" (p. 334). These sets of problems or problem types need not be hidden from students. The key, as suggested by practice 8, is to draw these problem types out and make them explicit to students by asking students to look for and express similarities (or differences) between the reasoning they are doing on their current problem and the reasoning they did on past problems. Questions such as "How is this problem similar to the last problem?" or "Why can't we take the same approach to this problem as we did on the other problem?" can help to elicit these sorts of ideas. They may also help send the message that the work of mathematics does not cease when an answer is reached. One can always continue to think by looking for generalities, connections, or distinctions.

Connections to Other Practices

It is perhaps surprising that looking for and expressing regularity in repeated reasoning is included as the last of the key mathematical practices in the *Common Core State Standards* because it can naturally serve as a starting point for a wide range of mathematical activity—for example, communication, representation, reasoning, and justification. Communication comes into play as soon as a mathematical regularity needs to be expressed. Not only must the expression of a regularity help the entire classroom community come together around the phenomenon, but it must also accurately communicate the conditions under which this regularity is expected to hold. In particular, when making general statements, it is necessary to attend to the precision (practice 6) of those statements—for example, does a relationship hold for all even numbers, for all natural numbers, for all integers, for all real numbers? By clearly communicating the conditions of the generalization, this sets the stage for the construction and critique of arguments (practice 3).

The description of this practice of looking for and expressing regularity also highlights its connection to problem solving (practice 1). First, as was discussed at the opening of this chapter, practice 8 in some ways describes what students should be thinking about as they work through problems or exercises, especially those that involve a repetition of reasoning. Students should be looking for commonalities, shortcuts, generalizations, and patterns in their reasoning, and this should be viewed not as separate to the problem solving but as an integral part of it. Second, the awareness of problem types can be beneficial to students in their problem solving because it can allow them to leverage past experiences in productive ways, such as identifying a subgoal within a problem that is similar to something they have already solved. And finally, the description of this practice includes several aspects of *metacognition*, or a person's awareness and thinking about his or her own thinking. To "maintain oversight of the process" and "continually evaluate the reasonableness of their intermediate results" (CCSSI 2010, p. 8) are clear examples of metacognition. For several decades, such metacognitive acts have been recognized as an important component of mathematical thinking, but they have typically been classified under the heading of problem solving because they play such an important role in the success of the problem solver (e.g., Schoenfeld 1985). By explicitly including metacognition in the practice of looking for and expressing regularity, the *Common Core State Standards* have intimately tied this practice to problem solving. Metacognition also ties this practice to the previous practice of looking for and making use of structure (practice 7). That practice notes that proficient students can "step back for an overview and shift perspective" (CCSSI 2010, p. 8), which is also a metacognitive act.

Another mathematical practice that has implications with respect to looking for and expressing regularity is the appropriate use of tools (practice 5). In some situations, the use of tools such as a graphing calculator or spreadsheet software may allow students to enact a large number of repetitions, thus exposing a regularity that they can explore further. Moreover, the use of these tools may free up mental space for students so that they can attend to the overarching regularities rather than the individual instances. It is also possible, however, that the use of tools may make the most important regularities invisible. For example, a pattern may lie within the process of calculating the individual instances instead of in the outputs of those calculations, so having a tool simply display the outputs is counterproductive. In such cases, it is advisable to have students do the calculations, but not mindlessly (like a computer). Students should be doing the calculations and actively looking for regularities in their repeated reasoning. Therefore, it is often crucial that the use of tools be carefully considered in relation to the practice of looking for and expressing regularity in repeated reasoning.

Although there are many connections between this practice and other key mathematical practices, one may wonder what the distinction is between the practices of looking for and expressing regularity in repeated reasoning and looking for and making use of structure. As described above, the current practice deals with the identification of a pattern across the products of reasoning (e.g., noticing that a relationship that holds for three particular cases will, in fact, hold for infinitely many cases) or with the identification of patterns in the reasoning itself (e.g., noticing that the Pythagorean theorem can be used in many different situations whenever an unknown side of a right triangle needs to be determined). Looking for and making use of structure, on the other hand, deals with the way in which mathematical objects are perceived and parsed—it can occur with a single object, with no need for repetition or a focus across instances. Roughly speaking, looking for and expressing regularity in repeated reasoning is a more global practice whereas looking for and making use of structure is more local. With that being said, however, mathematical structures often hold across multiple instances, so perceiving structure in the objects often comes hand in hand with perceiving regularity in reasoning around those objects. Thus, it is important to remind ourselves that drawing a hard line between practices is not as important as engaging students in these mathematical practices, whichever they may be.

Classroom Examples

To illustrate the various dimensions of practice 8, this section includes several practical examples that are somewhat shorter than examples in other chapters. At the elementary level, sample questions and tasks around number and operations are presented that can promote looking for and expressing regularity. A short classroom excerpt is also included related to geometric formulas as well as a discussion of the potential benefit of misgeneralizations. At the secondary level, two classroom excerpts are shared—one involving connections across inverse operations and the other involving a generalization of reasoning in trigonometry. In between these excerpts is a sample task that may be used in a geometry class to promote the recognition of regularities in proof arguments.

Elementary Grades Vignettes

Number patterns. In the early grades, numbers are a primary object of reasoning with which students can look for and express regularities and patterns. There are countless number patterning tasks that can provide opportunities to cultivate this mathematical practice in students. Below are a few that are designed to bring forth mathematical regularities so that students can talk about them and perhaps use them as opportunities to engage in other practices as well.

- Choose two even numbers and add them together. Repeat this until you notice a pattern in the sums. Do you think the pattern works for any pair of even numbers?
- Choose an even number and multiply it by an odd number. Repeat this until you notice a pattern in the products. Do you think the pattern works for the product of any even number and any odd number?
- Choose three prime numbers, three square numbers, and three other numbers (not prime or square). Find all the factors of each number that you chose. Do you notice any patterns about the number of factors for each kind of number? If you find a pattern, will it always be true for that kind of number?

Young students work frequently with numbers and, in many cases, repeat reasoning over and over again (e.g., for practice). Such repetitions can be dangerous if done mindlessly, but can also be especially beneficial for students if they use them as opportunities to look for patterns, make generalizations, and then communicate with one another about those generalizations. The sample questions above only scratch the surface of the generalizations that students can look for and express in the area of whole numbers. The important point is not the particular number pattern that is found but developing the habit of looking for and expressing them.

Types of subtraction. In conjunction with number, operations are another foundational topic in the early grades. Whereas the tasks above involved properties of numbers with respect to certain operations, the task in figure 8.1 takes a single operation—subtraction—as the explicit object of reflection.

What is the same about the four problems below? What is different?	
(A) A child has 7 pieces of candy. He gives 3 pieces to a friend. How many pieces does he still have for himself?	(B) A person with $12 buys an item for $8. How much money does the person have left?
(C) Packet A contains 9 pieces of candy. Packet B contains 5 pieces. How many more pieces does A have than B?	(D) If one job pays $11 and another pays $7, how much more does the first job pay than the second?

Fig. 8.1. Task related to subtraction

This sample task can provide an opportunity to think about types of subtraction problems such as take-away problems (A and B) or comparison problems (C and D). The types can be illuminated by having students represent the subtraction with physical manipulatives, drawing attention away from the surface characteristics and revealing the underlying structure. Research has found that a focus on the regularities in problems (and in students' approaches to those problems) such as subtraction problems has a strong positive relationship to students' learning. Furthermore, by recognizing that a variety of different situations can all be represented by subtraction, students can abstract a multitude of concrete situations into a single mathematical operation. This abstraction is itself a key mathematical process.

Thinking about formulas. Formulas in mathematics are often the source of rote drills and procedural exercises, but they can also provide opportunities to look for regularities in repeated reasoning, and formulas themselves can be viewed as expressions of those regularities. The following interaction comes from an elementary geometry lesson.

Ms. Kipling: Now let's look back at all of these squares. We've found five different perimeters. The squares were different sizes and had different perimeters, different lengths around the outside, but what did we do for each one of the squares to find the perimeter? Kylie?

Kylie: Added the sides.

Ms. Kipling: Yeah, even though the sides were different numbers for the different squares, what we did was the same—we added up the sides. And because these are squares, what else was the same?

Students: The sides.

Ms. Kipling: For each square, we're not only adding up all the sides but we're adding the same number each time because the four sides of a square are the same length. So here we had 3 plus 3 plus 3 plus 3 [*points to first square on the board*], here we had 7 plus 7 plus 7 plus 7 [*points to second square*], and the same here and here.

Aisha: That's like times 4.

Ms. Kipling: Ah, does everyone see what Aisha's thinking about. Aisha, can you say a bit more about what you mean?

Aisha: Well, the 3 plus 3 plus 3 plus 3 is just like 3 times 4, so I just did it as 3 times 4 to get 12.

Ms. Kipling: Aha. Because we're adding up the same number four times, we can also think of that as 3 times 4 or as 7 times 4 for the other square. So what I want everyone to do right now is, in your math notebooks, write down what it is that we do to find the perimeter of a square. You can write it down in whatever way makes sense to you. And remember, for any square, you're going to have four sides with the same length.

Essentially every formula can be viewed as an encapsulation of a process of reasoning that has been and will be repeated. By expressing it as a formula, work can proceed efficiently, but it is valuable for students to have experience with the reasoning that underlies particular formulas (e.g., one can imagine middle school students having an interaction similar to that above but focusing instead on the rule for dividing by fractions or high school students discussing the process of completing the square and the development of the quadratic formula) and to have experience generalizing from their repeated reasoning to develop a formula. NCTM (2000), in its *Principles and Standards*, notes that "many students will naturally seek a formula" (p. 224), but the teacher plays an important role in promoting such thinking and in guiding the collective expressions of the formulas.

The example above dealt with a regularity in reasoning that led to a formula for finding the perimeter of a square given its side length. But there are also instances of regularity *across* formulas. For example, when learning about volume, a teacher could engage students in looking for and expressing the regularity across volume formulas such as the formulas for cylinders and prisms—namely, determining the area of the base (in square units) and then multiplying this by the height (yielding cubic units). This general notion could even be viewed as a three-dimensional analog of the two-dimensional area formulas for rectangles and general parallelograms.

Generalization and misgeneralization. The practice of looking for and expressing regularity in repeated reasoning often takes the form of students generalizing mathematical patterns and relationships that they notice as they work. Tasks and classroom activities can be designed to lead students to discover and then express important generalizations such as the commutativity of addition and multiplication. As students become enthralled with finding and expressing general properties (it is fun to do, isn't it?), it is important for teachers to keep a careful eye out for misgeneralizations, which can nonetheless provide important learning opportunities. For example, it is important that students realize commutativity does not extend to subtraction or division.

A common misgeneralization for students in the elementary grades is that multiplication always makes things bigger. This idea stems from students' typical early experiences multiplying whole numbers exclusively, and they mistakenly think of this as a property of multiplication rather than a property of multiplication of whole numbers. The fact that the misgeneralization exists, however, does not mean that the underpinning idea should be completely discarded. The mathematical practice of looking for and expressing regularity implies that students should be encouraged to explore and articulate the boundaries of their generalizations. In the case of multiplication, they should be given a chance to understand the special role of 1 with respect to the operation of multiplication and the effects of multiplying by positive numbers less than 1 and greater than 1 (as well as multiplying by negative numbers). The boundaries of one generalization are often a good place to look for other generalizations.

Middle and High School Vignettes

Inverse operations. The following secondary example takes place near the beginning of a unit on logarithms, after logarithms have been introduced and defined but before students are very comfortable with them. The teacher, Ms. Rhodes, is guiding her students into making a connection between reasoning with logarithms and reasoning with other well-known operations.

Ms. Rhodes: In this equation ($5^x = 29$), the x variable that we are trying to solve for is in the exponent of the 5. So 5 has been raised to the x power, and we want to undo that so that we can see what x is by itself. And really, this is a one-step problem very similar to what you've been doing for a long time. Imagine that you have 5 plus x (*writes* $5 + x = 29$ *on the board*), how would you undo that to find x?

Student: Subtract 5.

Ms. Rhodes: Right, subtract 5 from both sides leaving x by itself. What if we had 5 times x (*writes* $5x = 29$ *on the board*)?

Student: Divide.

Ms. Rhodes: Division undoes the multiplication on the left and leaves you with your answer on the right. What about x divided by 5 (*writes* $\frac{x}{5} = 29$ *on the board*)? What can we do in that case?

Student: You can multiply both sides by 5.

Ms. Rhodes: Are people seeing a pattern here? One more. What if we had x to the fifth power [*Writes* $x^5 = 29$ *on the board*]? Can we undo this one? [*pause*] In this case, we can take the fifth root of both sides or raise both sides to the one-fifth power. So, thinking about all of these, what would you say is a general strategy for solving for a variable when there is an operation happening to it?

Student: You can do the opposite operation to get x by itself.

Ms. Rhodes: That's right. In all these cases, the inverse operation undoes whatever it was that happened to x so that we can clearly see what x is [*see fig. 8.2*].

$$5 + x - 5 = 29 - 5 \qquad \frac{5x}{5} = \frac{29}{5} \qquad 5\left(\frac{x}{5}\right) = (29) \cdot 5 \qquad \sqrt[5]{x^5} = \sqrt[5]{29}$$

Fig. 8.2. Connecting reasoning with logarithms and reasoning with other operations

Ms. Rhodes: Now, let's bring this pattern back to our original problem, 5 to the x power equals 29. How can we undo this exponentiation? What is the inverse operation of exponentiation?

Students: Logarithm.

Ms. Rhodes: Logarithms are our tool for solving this kind of equation [*see fig. 8.3*].

$$\log_5 5^x = \log_5 29$$

Figure 8.3. Completing the connection

In this example, the teacher is addressing not only the content standards related to logarithms but also the expectation that students "understand and use the inverse relationships" of various operations (NCTM 2000, p. 214; see also F-BF.5 in CCSSM [2010]). The teacher in the excerpt leads the intellectual work, but it would also be worthwhile to have students take the lead in communicating or working with these ideas, perhaps by having them solve and discuss the variety of equations presented in the example or by asking them to write a few sentences about the relationship between logarithms and exponents and how this compares with other pairs of operations. Such activities not only draw attention to regularities in mathematical reasoning, but also set the stage for higher levels of mathematics such as abstract algebra, where the concept of a *group* is essentially the abstraction of the pattern noted by Ms. Rhodes.

Proof types. The task in figure 8.4 could be used in a high school geometry course or a modification could be used in any course that involves writing proofs. Alternatively, instead of a written task, this idea of patterns across proofs or proof types could be a topic of discussion with students.

A task such as this creates opportunities for students to do more than prove mathematical statements; they can analyze mathematical reasoning and look for patterns or regularities in that reasoning. Students may begin by focusing on surface features of the proofs such as their layout or length, but if given the chance, they can dig deeper to uncover such patterns as the use of corresponding parts of congruent figures or the dissection of a polynomial into triangles. Furthermore, it is important for students to be readers of proofs and mathematical justifications, not only producers. This task gives students opportunities to read proofs and think about or discuss them. Providing such opportunities to step out of the mathematical work and look at it for its own sake is an important part of cultivating the mathematical practices in a classroom. A similar function can be served by going over homework in ways that encourage students to not only consider individual problems but also look across problems for any regularities in their repeated reasoning.

Repeated reasoning on the unit circle. The following classroom example from a trigonometry unit involves a teacher helping students realize that the reasoning involved in proving a relationship for a specific angle can be repeated for other angles as well, yielding a general result. Noticing this regularity in reasoning—that an argument can be repeated for an arbitrary element of a class of mathematical objects—is a key to understanding and being able to develop mathematical proofs and general justifications (see practice 3).

The students have been working to determine the value of $\sin^2(30°) + \cos^2(30°)$. Several students used $\sin(30°) = 1/2$ and $\cos(30°) = \sqrt{3}/2$ to solve the problem, but now the teacher, Mr. Johns, is going to call on a student who approached the problem in a different way.

Mr. Johns:	Marco, would you mind coming up to explain how you thought about this problem?
Marco:	[*Comes to the front board and draws the triangle in fig. 8.5.*] So I thought about sine of 30 degrees as *a* over *c*. And cosine is *b* over *c*. So then I squared those because we have to add up the squares [*writes on board, see fig. 8.5*]. And then I just simplified it to 1. I got 1 as the answer.

Compare and contrast the mathematical proofs below.

CLAIM: If the two diagonals of a convex quadrilateral bisect one another, then its opposite sides are of equal length.

PROOF. Let *ABCD* be a convex quadrilateral with *E* the intersection point of diagonals *AC* and *BD*.

Conclusions	Justifications
$AE = EC$, $BE = ED$	Hypothesis
$\angle AEB = \angle CED$	Vertical angles
$\triangle AEB = \triangle CED$	Side-angle-side
$AB = CD$	Corresponding parts of congruent triangles

Q.E.D.

CLAIM: If *ABC* is a triangle with $AB = AC$, then the bisector of $\angle A$ is the perpendicular bisector of side *BC*.

PROOF: Let *ABC* be an arbitrary isosceles triangle with *A* the vertex where the congruent sides meet, that is, $AB = AC$. Let *D* be the point on side BC such that *AD* bisects $\angle CAB$. This means that $\angle CAD = \angle BAD$. We also know that $AD = AD$ by reflexivity, so $\angle CAD = \angle BAD$ by the side-angle-side congruence condition. Therefore, $CD = BD$ and $\angle CDA = \angle BDA$ because these are corresponding parts of the congruent triangles. This means *D* is the midpoint of side *BC* and *AD* is perpendicular to *BC*. By definition, then, *AD* is the perpendicular bisector of *BC*.

Q.E.D.

CLAIM: If ABCD is a quadrilateral, then $\angle A + \angle B + \angle C + \angle D = 360°$.

PROOF: Let *ABCD* be any quadrilateral with diagonal *AC* drawn in.

Conclusions	Justifications
$\angle BAC + \angle CAD = \angle BAD$	Angle addition
$\angle B = \angle B$	Reflexivity
$\angle BCA + \angle ACD = \angle BCD$	Angle addition
$\angle D = \angle D$	Reflexivity
interior angle sum of $\triangle ABC$ and $\triangle ACD$ equals interior angle sum of *ABCD*	Sums of equals are equal
interior angle sum of $\triangle ABC$ and $\triangle ACD$ together equals 360°	Triangle sum theorem
interior angle sum of *ABCD* equals 360°	Transitive property of equality

Q.E.D.

CLAIM: The interior angles of a quadrilateral sum to 360 degrees.

PROOF: Let *ABCD* be a quadrilateral and let E be a point in the interior of *ABCD*. By connecting *A, B, C,* and *D* to *E*, four triangles are formed. By the triangle sum theorem, each of these triangles has interior angles that add up to 180° and so, all together, the interior angles sum to 720°. Because $\angle AEB$, $\angle BEC$, $\angle CED$, and $\angle DEA$ are interior angles of the triangles but do not constitute interior angles of *ABCD*, we must subtract the measures of these angles from the total to determine the interior angle sum of *ABCD*. These four angles form two complementary pairs and so, by the definition of complementary angles, equal 360° in sum. Therefore, the interior angles of *ABCD* sum to 720° – 360° = 360°.

Q.E.D.

Fig. 8.4. Task for seeing patterns across proofs

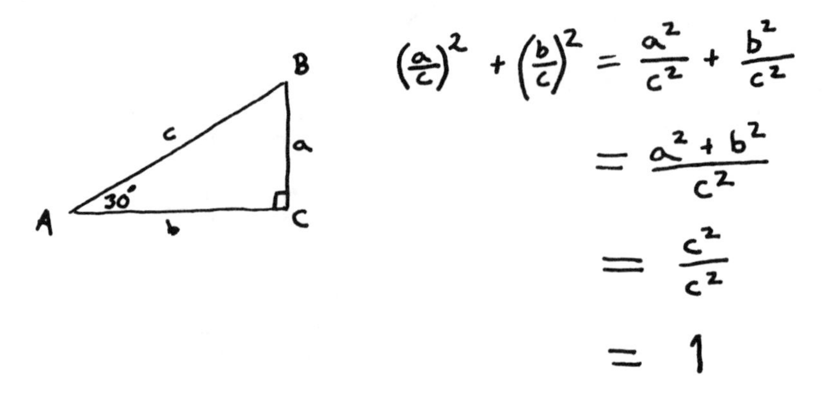

Fig. 8.5. Marco's solution

Mr. Johns: Before you sit down, Marco, I wanted to ask about this step [*points to the substitution of* c^2 *for* $a^2 + b^2$]. What's going on here?

Marco: Well, *a*-squared plus *b*-squared is equal to *c*-squared, so I just put *c*-squared there instead.

Mr. Johns: [*to the class*] Now, it's not always the case that *a*-squared plus *b*-squared equals *c*-squared, but it is true here because why? Chloe?

Chloe: Because it's a right triangle.

Mr. Johns: Yeah, the fact that we're working with a right triangle allows us to do this substitution, and from there it comes out as 1. Now I want you all to try something. I want everyone to take about a minute to try Marco's method to figure out a new problem [*writes on board* $\sin^2(75°) + \cos^2(75°)$]. Try this same method. [*Students work for several moments as Mr. Johns draws a new triangle on the board.*] OK, what did you find?

Students: It's the same.

Mr. Johns: What do you mean, "It's the same"?

Jacque: I did exactly the same thing Marco did, and it ended up being 1. It was the same for 75 degrees.

Mr. Johns: So what you're saying is that, with this new triangle, we still have sine of the angle equaling *a* over *c* and cosine equaling *b* over *c* [*writes on board*]. So the rest simplifies exactly the same way to be 1. In fact, does it matter at all whether this angle is 30 or 75 or anything else?

Mr. Johns continued in this lesson by introducing the general identity $\sin^2 \theta + \cos^2 \theta = 1$ for any angle θ. He encourages students to connect the repeated reasoning of the 30° and 75° instances to the justification for the identity. By having students work through multiple instances, he laid the groundwork for their noticing that the line of reasoning actually holds independent of the specific acute angles, though it does depend on the presence of the right angle. The mathematical content in this example is F-TF.8 in CCSSM (CCSSI 2010, p. 71), but it is the mathematical practice of looking for regularities in repeated reasoning and developing that regularity into a justification that helped the lesson come to life.

Conclusion

The examples above include tasks that provide opportunities for students to attend to regularities in their reasoning as well as classroom interactions. These examples highlight how a teacher might work to bring out this practice of looking for and expressing those regularities. Although efforts toward this practice can be carefully planned, we should also keep an eye out for patterns or regularities that students notice even when we are not expecting them. A goal is for students to make a mental habit out of looking for these regularities, and if we succeed in this goal, students may then surprise us in what they notice and the mathematical observations they make.

Resources

Books and Book Chapters

This book in the Essential Understanding series is based on the central role of generalization in the development of students' algebraic thinking skills, and generalization is a key component of the practice of looking for and expressing regularity in repeated reasoning.

- Blanton, M., L. Levi, T. Crites, B. Dougherty, and R. M. Zbiek. *Developing Essential Understanding of Algebraic Thinking for Teaching Mathematics in Grades 3–5.* Reston, Va: National Council of Teachers of Mathematics, 2011.

This chapter discusses the process of recognizing visual patterns and expressing them in various ways, which can provide a foundation for further mathematical reasoning.

- Billings, E. M. "Exploring Generalization through Pictorial Growth Patterns." In *Algebra and Algebraic Thinking in School Mathematics.* 2008 Yearbook of the National Council of Teachers of Mathematics (NCTM), edited by C. E. Greenes and R. Rubenstein (vol. 70, pp. 279–93). Reston, Va: NCTM, 2008.

Journal Articles

This article refers to virtual manipulatives that can be used to incorporate pattern recognition and algebraic thinking into the elementary grades. Links to several practices, including looking for and expressing regularity in repeated reasoning, can be made.

- Moyer-Packenham, P. "Investigations: Using Virtual Manipulatives to Investigate Patterns and Generate Rules in Algebra." *Teaching Children Mathematics* 11, no. 8 (2005): 437–40.

This article deals with geometric problems that teachers can use to have their students notice patterns, make predictions, and express generalizations algebraically.

- Beigie, D. "The Leap from Patterns to Formulas." *Mathematics Teaching in the Middle School* 16, no. 6 (2011): 328–32.

This article describes generalization and justification processes of middle school students and includes sample problems that may be useful to teachers at this level in cultivating the practice of looking for and expressing regularity in repeated reasoning.

- Lannin, J. "Developing Algebraic Reasoning through Generalization." *Mathematics Teaching in the Middle School* 8, no. 7 (2003): 342–46.

This article discusses students who generalized from pattern sequences and the pedagogical moves that supported this practice. The expressions of the patterns culminated in algebraic models (see practice 4).

- Rivera, F. "Connecting Research to Teaching Generalization." *Mathematics Teacher* 101, no. 1 (2007): 69–72.

References

Blanton, M. L., L. Levi, T. Crites, B. Dougherty, and R. M. Zbiek. *Developing Essential Understanding of Algebraic Thinking for Teaching Mathematics in Grades 3–5.* Reston, Va: National Council of Teachers of Mathematics, 2011.

Carpenter, T. P., E. Fennema, P. L. Peterson, C.-P. Chiang, and M. Loef. "Using Knowledge of Children's Mathematics Thinking in Classroom Teaching: An Experimental Study." *American Educational Research Journal* 26 (1989): 499–531.

Common Core State Standards Initiative (CCSSI). *Common Core State Standards for mathematics.* Washington, DC: National Governors Association Center for Best Practices and the Council of Chief State School Officers, 2010. www.corestandards.org/assets/CCSI_Math%20Standards.pdf.

Ellis, A. B. "Generalizing-Promoting Actions: How Classroom Collaborations Can Support Students' Mathematical Generalizations." *Journal for Research in Mathematics Education* 42 (2011): 308–45.

Jitendra, A. K., J. R. Star, K. Starosta, J. M. Leh, S. Sood, G. Caskie, et al. "Improving Seventh-Grade Students' Learning of Ratio and Proportion: The Role of Schema-Based Instruction." *Contemporary Educational Psychology* 34 (2009): 250–64.

National Council of Teachers of Mathematics (NCTM). *Principles and Standards for School Mathematics.* Reston, Va: NCTM, 2000.

National Research Council (NRC). *Adding It Up: Helping Children Learn Mathematics.* Edited by J. Kilpatrick, J. Swafford, and B. Findell. Washington, D.C.: National Academies Press, 2001.

Schoenfeld, A. H. *Mathematical Problem Solving.* Orlando, FL: Academic Press, 1985.

Connecting the Practices

THE INCLUSION OF THE EIGHT STANDARDS FOR MATHEMATICAL PRACTICE in the *Common Core State Standards for Mathematics* (CCSSM; CCSSI 2010) helps fully represent what it means to *do mathematics*. However, the Standards for Mathematical Practice are not without flaws. Some might critique the brevity of the descriptions of the practices, and the preceding chapters are an attempt to address this critique. Others might critique the presentation of the practices as eight discrete entities, which has the potential to emphasize a separation between the practices rather than unity or connections between them. Therefore, we would like to conclude this book by considering some of the key links between practices.

Practice 1, making sense of problems and persevering in solving them, relates to many of the other practices, as problem solving is at the heart of mathematical activity. Indeed, as students work to solve an authentic problem, they may use quantitative reasoning skills (practice 2) or various tools (practice 5) to make progress in their solution, and within their work they may look for structure (practice 7) and regularity in reasoning (practice 8). Furthermore, students may be expected to explain and justify their solution rather than simply arrive at an answer, and explaining one's thinking is likely to involve the construction of an argument (practice 3) and an attention to precision (practice 6) in the act of communication. It may also be the case that the problem at hand necessitates that students generate and use a mathematical model (practice 4). In other words, students who engage in the first Standard for Mathematical Practice are likely to be employing the other practices as well.

Another mathematical practice that tends to involve several other practices at the same time is constructing viable arguments and critiquing the reasoning of others (practice 3). The germ of a mathematical argument is the claim or conjecture that fleshes out a mathematical idea or relationship. Looking for and expressing regularity in reasoning (practice 8) is a means by which students formulate conjectures and can serve as a catalyst for reasoning and proving. In constructing or evaluating the body of an argument, depending upon the mathematical context, students may need to apply quantitative reasoning (practice 2) or rely on inherent mathematical structure (practice 7). Mathematical modeling (practice 4), particularly when one validates the appropriateness of a mathematical model, involves constructing viable arguments to convince others of the strength of the model. Finally, when students communicate their mathematical arguments, strategically using tools (practice 5) and representations of mathematical thinking can support others in understanding and critiquing their arguments. Likewise, attending to precision (practice 6) of language and communication is a necessary aspect of constructing valid mathematical arguments.

The preceding paragraphs described some key practices that often entail other practices at the same time. There are also some pairs of practices that tend to occur together and may even be difficult to distinguish in the classroom. One example of this is the practice of constructing viable arguments and critiquing the reasoning of others (practice 3) and the practice of attending to precision in mathematical language (practice 6). As students express their reasoning to form an argument, whether verbally, in writing, or through the use of symbols or diagrams, the viability and validity of their argument is tied up with the precision of their communication. Moreover, if students are critiquing one another's reasoning, these critiques may be based on imprecision in the argument.

A final connection that we mention here is between looking for and using structure (practice 7) and looking for and expressing regularity in repeated reasoning (practice 8). The former involves students recognizing the structure of the ideas or mathematical entities they are working with and using that structure to gain insight into the mathematics or to make progress toward a solution. Mathematical structures, however, often recur and if students are using these *structures* in their reasoning then it is likely that students will also have opportunities to notice *regularities* in their reasoning. These practices related to structure and regularity may often coalesce within processes of generalization. This push toward generalization and extending beyond a particular situation also ties in with abstract reasoning (practice 2) and may build toward justification (practice 3) of the generalized claims.

It would be difficult, and probably unwise, to try to enact a single practice in isolation from others. If students are engaging in rich mathematical activity, they are most likely engaging in several of the mathematical practices. In *Adding It Up* (NRC 2001), the five strands of mathematical proficiency were depicted and discussed as an interwoven braid; in a similar way, there is a symbiosis among many of the Standards for Mathematical Practice. As teachers implement meaningful activities related to one or more of the practices, students are likely to be simultaneously engaged in other practices as well. Furthermore, as teachers navigating the "multidimensional terrain" (Lampert 2001, p. 50) that is mathematics education, our focus can be on making sure that these practices lead to conceptual understanding of the content standards and that students themselves are aware of the nature of their activity. If we notice students engaging in one or more of the practices, it may be worthwhile to step back with them to point out those productive forms of mathematical activity. This explicitness can increase the likelihood that the practices will arise again in the future, and we can help students realize that the practices they are engaging in are characteristic of what it means to *do mathematics*.

References

Common Core State Standards Initiative (CCSSI). *Common Core State Standards for Mathematics.* Washington, D.C.: National Governors Association Center for Best Practices and the Council of Chief State School Officers, 2010. http://www.corestandards.org/assets/CCSI_Math%20Standards.pdf.

National Research Council (NRC). *Adding It Up: Helping Children Learn Mathematics.* Edited by J. Kilpatrick, J. Swafford, and B. Findell. Washington, DC: National Academies Press, 2001.

Lampert, M. *Teaching Problems and the Problems of Teaching.* New Haven, Conn.: Yale University Press, 2001.